The Problem With Suffering

AND THE SCIENCE OF HOPE

Also by Roddy Carter

BodyWHealth: Journey to Abundance

Sunset Lessons: Reflections on Light and Love from the Darkest of Places

Fireside Wisdom: Conversations to Inspire Personal Mastery

Becoming Unstoppable: Your Neurocentric Coaching Guide to Achieving Unstoppable Success

The Serpent Within: Navigating Fear to Restore Inner Harmony

***Conscious Pathways* Series**

The Problem With Anger: And How to Solve It

The Problem With Growing Up: Confronting Imposter Syndrome

***Little Book* Series**

Unstoppable Starts Here: A Short Guide to Mastering Your Brain and Unlocking Your Life

Greatness Awakened: A Short Guide to Mastering Your Complex Brain and Unlocking Your Best SELF

***Unstoppable You* Online Courses**

Unstoppable You

Unstoppable You Business

Unleash Unstoppable

Unleash Success

Visit www.roddycarter.com for the full collection of Roddy's writing, including new and forthcoming titles.

The Problem With Suffering

AND THE SCIENCE OF HOPE

by Roddy Carter, MD

Illustrated by Hannah Mae Silcock

Aquila Life Science Press
La Jolla, California

FIRST AQUILA LIFE SCIENCE PRESS EDITION, MAY 2026
Published by Aquila Life Science, LLC, La Jolla, CA

THE PROBLEM WITH SUFFERING.

Illustration by Hannah Mae Silcock.
Editing and book design by Sarah Dawson, WordPlay Editing.

ISBN: 979-8-9895751-7-6

Printed in the United States of America

For Janet, who suffered greatly.

And for Hope, who lives in all of us.

ACKNOWLEDGEMENTS

Thank you to all who have invited me into your sacred space when suffering has touched your life.

Thank you also to three extraordinary partners on my literary journey.

Sarah Dawson, once again, you have forged the balanced integration of complexity and clarity. Your unmatched editorial sensitivity and fierce commitment to truth-telling have helped shape not only the words but the soul of the work. You help me say exactly what I mean—always better than I could have said it alone. Our partnership is one of the most consistent and cherished pillars of my creative life.

Hannah Mae Silcock, your illustrations continue to breathe life into the seen and unseen spaces of this work. With your gift for visual storytelling, you make the science more human and the humanity more vivid. Your art enlivens the emotional texture of this work. Thank you for bringing soul to the science. I'm grateful to be on this journey with you, again.

Glynis Carter, your quiet brilliance continues to power the scientific depth and accuracy of my writing. With rigor and intuition, you anchor the message in credibility. Thank you for your intellectual generosity, precision, and partnership that elevate everything we craft together.

PREFACE

I have lived in the presence of suffering.

I have witnessed it in the eyes of loved ones, patients, and clients. I have felt it in my own body, mind, and—perhaps most deeply—spirit. It has shadowed moments of joy, shaped seasons of my life, and called me into work that I now recognize as my life's calling.

This book is the product of that calling, a response to suffering's grip on the people I love and its gnawing presence in my own life. It is the fruit of decades of clinical practice, scientific inquiry, and lived experience. But more than that, it is a deeply personal offering. A field guide. A reckoning. A conversation.

I do not pretend to fully understand suffering. I do not write from a place of triumph or mastery over it. I have not conquered it in myself, and I certainly do not claim to have eliminated it in others. This is not a manifesto of certainty or a manual of miracles. Rather, it is a reflection of one man's ongoing journey—drawn from neuroscience, psychology, philosophy, and personal observation—and offered in the sincere hope that it will be useful to you on your own path.

Please know that, in my underlying thesis, I do not minimize your pain. I am no stranger to the weight of suffering.

I come to this work not as a detached expert but as a student. As a scientist with deep familiarity with the human brain, I have become convinced that we suffer far more than we need to—not because we are weak, but because we are untrained. We have inherited biology that makes us vulnerable to suffering, and we often lack the knowledge, tools, or support to manage it wisely.

This book invites you to challenge that predicament, to look more closely at what suffering is, where it comes from, and how it can be softened...or transformed altogether. I hope you will take it as a provocation, not to deny suffering, but to explore it with courage, clarity, and compassion. And where suffering proves unavoidable, I hope you will meet it as your teacher—one who reveals who you are, who you are not, and how you might live with greater peace, purpose, and joy.

Above all, this book is an invitation to hope.

If you are suffering now, or fear that suffering may visit you or someone you love in the future, I pray that you will find hope—perhaps Hope herself will find you—within these pages. And if you are walking alongside someone who suffers, may this book give you language, light, and tools to be present in that sacred space.

In the end, I cannot tell you to what extent suffering is either good or bad. I can't prove that it serves a necessary role in the arc of every life. But I do know this: It is a miserable condition to endure. And for that reason alone, I have committed myself to easing its grip, first in myself and then, wherever possible, in others.

That work—my deepest calling—lives in these pages.

Welcome to *The Problem With Suffering: And the Science of Hope*.

CHAPTER 1

THE STORY OF SUFFERING

Hope was a curious little girl with a heart full of kindness. She loved exploring the world around her, asking questions and helping others whenever she could. Yet, despite her bright spirit, Hope often felt a deep sadness clouding her heart.

She didn't understand why she felt this way, but the weight of it followed her like a shadow.

Hope lived with her parents in a cozy house at the edge of town. She enjoyed her days at school, but her favorite spot was the local park with its towering trees, colorful flowers, and sparkling pond. It was a place where she could escape to reflect on her thoughts.

One afternoon, as Hope wandered through the park, she noticed an old woman sitting alone on a bench under a large oak tree.

The woman was short in stature, with deep, warm eyes and dark black hair flecked with gray streaks. A heavy blanket shrouded her stooped shoulders. She looked as though she was carrying a lifetime of suffering. Despite her sadness, the woman flashed a welcoming smile that reached deep into Hope's heart.

Intrigued, Hope walked over and spoke softly. "Hello. My name is Hope. May I sit with you, please?"

The woman hesitated for a moment, then nodded gently. "Of course, dear. My name is Janet."

Hope sat beside Janet, feeling a mix of curiosity and empathy. After a few moments, Hope spoke again. "You seem sad."

Janet sighed, her eyes softening at Hope's earnest concern. "I've had a long life, Hope, filled with both joy and sorrow. Sometimes, the weight of my memories can be overwhelming."

Seeing the suffering in Janet's eyes, Hope began to talk about the park, the weather, and other small, simple things. Her gentle, inquisitive nature made Janet feel comfortable and understood, something she hadn't felt in a long time.

As the afternoon turned to evening, Hope felt a growing connection with Janet. She wanted to learn more about her life and the reasons for her sadness. "Would you like to share your story with me?" Hope asked. "I would love to listen."

Janet looked at Hope. She could see that the young girl was genuinely interested. Taking a deep breath, Janet began to share her tale—opening a door to the past that had long been closed. Hope listened intently, with an open heart.

"I was very young when my life changed," Janet started. "I was only four years old when the war came. My father was a soldier, a brave man who left my mother and me to fight for peace. He never returned."

Janet paused, the quiet emphasizing the pain of that loss.

"The city was under attack, and my mother and I were sent away by train to a distant country. The journey was long and frightening. My mother tried to stay strong for me, but I could see the fear in her eyes, and when I lay against her chest at night, I heard her heart pounding."

"Not long after we reached our new home, my mother fell ill. There were no familiar faces, no family to turn to. Her condition worsened, and eventually, she died in a cold, white hospital bed. I was left alone."

Janet stopped, her voice breaking. The weight of the silence between her words was heavy, as if the air itself was holding its breath.

Hope felt the depth of Janet's isolation and fear, and her heart ached for the little girl Janet had been. She looked up into the face of her new friend, now wet with tears. Hope tenderly lifted her sleeve to wipe them away.

After a long pause, Janet's voice continued. "But then something miraculous happened: An older woman took me in. She was a warmhearted doctor, and though she had no children of her own, she treated me with affection that I had never known from a stranger.

"She gave me a stable, loving home. She was much older than my own mother, but she was wise and compassionate. She taught me so much about life. But, despite her kindness, the fear of being abandoned never left me. It was a shadow that followed me everywhere."

Hope could see the mixture of gratitude and sorrow in Janet's eyes as she spoke. "It was a constant struggle," Janet admitted. "I tried to move past the pain, but it always lingered, affecting my relationships and my happiness.

"I eventually married and had two children of my own. For a while, I thought I had escaped the shadows of my past."

Janet's voice grew softer.

"But the weight of my suffering eventually destroyed my marriage. The sadness I carried with me strained my relationships with my children."

Despite her obvious suffering, there was a resilience in Janet that Hope found inspiring. "But you never gave up," Hope said softly.

Janet smiled warmly. "No, I didn't give up. I learned that there's always hope, even in the darkest times. I leaned that you only experience shadows when there is a light shining somewhere. It's our job to find that light."

Moved by Janet's story, Hope felt a deep desire to learn more about suffering and how to overcome it. Since both were eager to continue their conversation, they returned to the park to sit and talk as often as they could.

One afternoon, Hope gathered her courage and spoke. "Janet, your story has touched me deeply. I want to understand more about suffering and how to escape it. Would you guide me on this journey?"

"It's a difficult path," Janet warned. "But if you're truly willing, I will share what I know."

Hope nodded eagerly. "Yes, I'm ready."

Janet smiled. "Very well. Let's begin."

They decided to revisit places from Janet's past that had shaped her life and her understanding of suffering.

Their first stop was the old railway station where Janet had arrived as a frightened child. As they walked through the deserted platforms, Janet recounted her memories. "This is where I first felt the true weight of my loss. I was just a little girl, and the world felt so big and overwhelming. The platform was filled with hundreds of other mothers and children fleeing the vile grip of war.

"Here, I learned my first lesson about suffering: It is a universal experience. Everyone has their own burdens to bear."

Next, they paused at the hospital where Janet lost her mother.

She described the cold white walls and the gaunt faces of the hardworking nurses in their starched uniforms who had wrestled Janet from her dying mother's arms. She shared how they had placed her into the care of a grumpy old clerk who smelled of cigarette smoke and dirt.

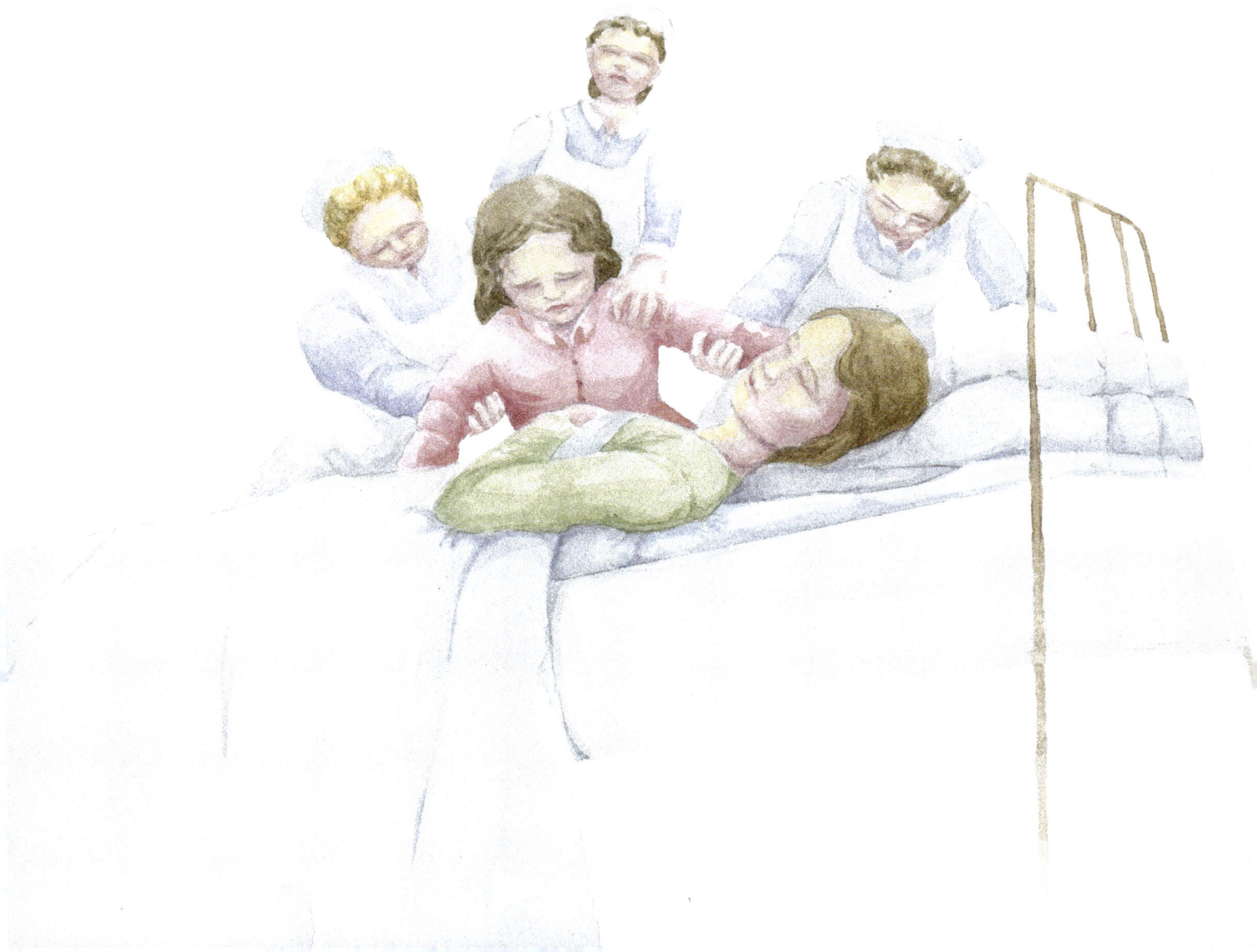

Janet described how this vision was seared into her memory. She could relive the moment as though she was still there. And her suffering mind took her back over and over again to that place, where time seemed to stand still, forever.

Hope listened intently. She could almost see the scared little girl Janet had been, standing helpless in this vast space.

"Not long ago," Janet whispered, "I realized that this image was no longer real. The little girl was no longer here, but her pain lived in my memory and consequently in my life, every day. My mind wouldn't let go of her. It fought to hold onto those painful images."

They sat quietly for a long time, contemplating the desperate scene and the revelation that Janet had stumbled upon.

After a while, they visited the small house where Janet had lived with her adoptive mother.

"This house was a sanctuary for me," Janet said. "It was here that I learned about love and kindness. My new mother was a doctor devoted to fighting against suffering. She taught me that, even in the darkest times, compassion can be a powerful healer.

"Although she had suffered in her own life, she never let it define her. Instead of mourning a childless marriage or the early death of her husband, she turned her pain into strength to help others."

Hope felt a surge of admiration for Janet's adoptive mother. "She sounds like a remarkable woman," Hope said softly.

"She was," Janet replied, her smile warm and reflective. "Her example taught me that, while we may not always be able to escape our suffering, we can transform it. In focusing on helping others and showing compassion, we can find purpose and healing."

Janet's voice grew more somber as she continued her story.

"When I got married and had two children of my own, I thought I had finally escaped my past. But my fear of abandonment and the sadness I carried with me strained my marriage. I was needy, very needy, and I held too tight. In fear of losing again, I crushed something very precious. And then, despite my best efforts, the weight of my suffering hurt my relationship with my children."

Desperately sad for Janet and afraid her own life would be similarly full of suffering, Hope asked, "Is there a way to overcome this shadow—to stop it from affecting the future?"

Janet looked at Hope, her deep, warm eyes filled with surprising strength. "Yes, there is," she answered firmly. "And I will help you find it."

As the sun set, Hope looked at Janet with gratitude. "Thank you for sharing your story with me," she said softly. "You've taught me so much."

"And you, my dear, have given me hope," Janet replied.

As time passed and their conversations continued, Hope began to see the world through a new lens. She noticed that suffering can be physical, emotional, or existential. She observed how suffering sometimes comes to people through natural disasters or the actions of others, and how it sometimes is caused by the people themselves. She understood that, while suffering is a universal experience, it is not insurmountable. With Janet's guidance, she learned that it is possible to transform suffering into strength and to find joy even in the midst of pain—maybe *especially* in the midst of pain.

Janet also taught Hope about the power of the human spirit. She reminded her that there is darkness only in the presence of light and that the greatest lesson is to turn away from the suffering to search for that light.

One afternoon, as the sun began to fade, Janet shared a recent transformative insight with Hope. "There was a time," Janet began, her voice calm and reflective, "when I felt completely overwhelmed by my suffering. It was shortly after my marriage ended. I was struggling to keep myself together for the sake of my children, but inside, I felt shattered.

"I sat in my bedroom, with my sons asleep in their own rooms, and I felt utterly lost. I felt the weight of my past pressing down on me, and I didn't know how to move forward. I was lost in the depths of my despair.

"Last week, I was reflecting on that moment of great darkness. It helped me to realize something very important. I wish I could have seen it at the time.

"You see, my boys were healthy. We were together, and I should have been happy. But instead, I was suffering...unnecessarily."

Janet paused, allowing Hope to fully grasp the gravity of her words.

"I understand now that all my suffering was rooted in my mind.

"I understand that everyone who suffers, no matter their situation, suffers internally.

"But I have also realized that the human spirit is indomitable. We have the power to choose how we respond to our suffering."

Hope's eyes widened as she absorbed Janet's words. "You mean we can control our suffering?"

"Yes," Janet replied.

"The problem with suffering is that ***we don't need to suffer***. But nobody teaches us this! When we realize that we are in charge of our own minds, we can manage our responses to painful experiences. When I realized that, the light of hope emerged from the shadows."

Hope felt a surge of inspiration. "So, what did you do next?" she asked.

"I have started to live differently," Janet explained. "Every morning, I take a few moments to gather my thoughts, reflecting on the positive aspects of my life. It wasn't easy at first, but I've noticed a shift in my perspective. I am beginning to appreciate the small joys and acts of kindness around me."

"Yesterday, my son came to me with a problem," she continued.

"In the past, I might have been too consumed by my own pain to offer him the support he needed. But this time, I remembered that I choose my own response. Rather than allowing my mind to focus on my recurring guilt and shame, I took a deep breath and listened to him with my whole heart. That simple act of being fully present made a world of difference. It strengthened our bond and showed me the power of my compassion when I put my own suffering aside."

Listening to Janet, Hope realized that it was possible to transform suffering into strength and that even the smallest actions can make a significant difference.

Janet continued, "We must be gentle with ourselves. We all make mistakes, and we all have moments of weakness. But if we can learn to forgive ourselves and others, we can release the heavy burdens of blame and shame that often accompany suffering."

Hope felt deep gratitude for Janet's wisdom. She began to apply these lessons in her own life, finding that they brought her resilience, peace, and joy.

One evening, as twilight settled over the park, Hope turned to Janet. "Thank you for sharing your journey with me, Janet. Your lessons have changed my life."

Janet's eyes glistened with a mixture of pride and tenderness. "Hope, you have reminded me of the power of connection and the importance of passing on our wisdom. It is you who has changed *my* life. Today, I understand that we can face any darkness with courage and compassion."

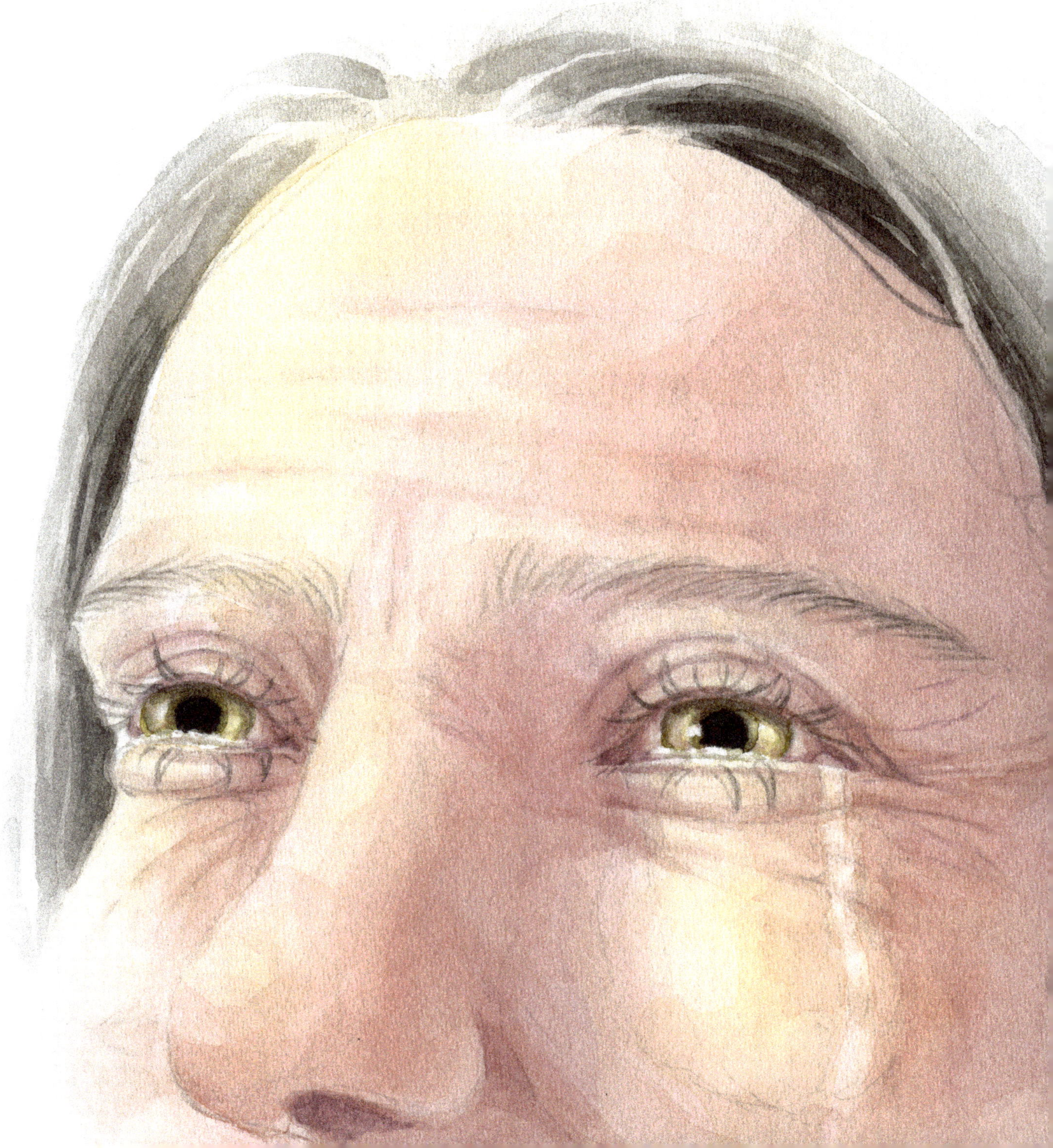

As she grew, Hope did not forget the lessons Janet had shared. They didn't spare her from struggle, but they gave her a way to meet it with courage, curiosity, and care. In time, she began to offer to others what she had learned—not as someone who had conquered suffering, but as someone who knew its terrain and could walk encouragingly beside those still finding their way.

Are you suffering unnecessarily?

And are you willing to invite Hope into your life, too?

AUTHOR'S NOTE

If the story of Janet feels intensely personal to you, it's no surprise to me, because her character is inspired by my mother.

In the quiet spaces of my childhood, I watched my parents each bear their own unique suffering, side by side but profoundly alone. My father, a man of steadfast faith, hid his pain behind a carefully constructed wall of stoicism. His suffering, though deeply felt, was locked away, leaving him in self-imposed isolation—a solitary confinement of the heart.

In contrast, my mother's pain was raw, visible to all who knew her. She had been abandoned to orphanhood at the age of four, and this wound lingered as an open ache throughout her life. Her grief was a palpable weight that often became too much for others to bear, eventually leaving her emotionally isolated despite her enormous capacity for deeply authentic love.

My parents' lives, though different, illustrate two familiar expressions of protracted human suffering: one hidden, the other openly worn, and both devastatingly painful. Despite their deep compassion for others, they remained tethered to an internal desolation that no amount of faith or love could seem to dissolve. Their suffering shaped not only their lives but the lives of their children.

More importantly, their suffering shaped my understanding of the human experience. Observing their anguish left me with deep questions about suffering—why it persists, how it shapes us, and whether we have the power to transform it. Their experiences, my own experiences, and a desire to know the answers to these questions ignited a flame deep within me to eradicate the pain of individual suffering. This has been my life's mission.

Suffering is a universal experience—a thread woven into the fabric of human life across all cultures, eras, and societies. Sometimes our suffering is the product of external events like loss, abandonment, or hardship. Other times, it seems to arise from an internal source, rooted in our perceptions, memories, and reactions.

The burden of suffering takes a toll not only on our minds but also on our bodies. It fractures lives, erects barriers between loved ones, and, as I saw with my parents, becomes a legacy passed down to children, subtly yet profoundly influencing future generations.

In the rest of this book, we will explore the complex nature of human suffering from a scientific perspective.

We will investigate the physiological, neurological, and psychological roots of suffering. We will examine how childhood experiences and trauma shape our reactions to suffering, often creating patterns that persist long after the original events have faded. We will discover that, while life presents inevitable challenges, the suffering that follows is often an internal reaction rather than an unavoidable reality.

Next, we will shift from understanding suffering to transforming it. Drawing on insights from neuroscience, psychology, and the contemplative traditions, we will explore practical methods for easing suffering. Through deliberate strategies, we will learn to apply the science of hope to reclaim authority over our inner experience and cultivate a life defined not by suffering but by peace and purpose.

This journey into the science of suffering is not an esoteric exercise but a warm and direct invitation to consider your own painful experiences with curiosity and compassion, to face your suffering, and ultimately to reimagine your relationship with it—and your life without it.

Here is an overview of what's to come:

- Chapter 2: A brief overview to the overarching hypothesis that we don't need to suffer, even through life's trials
- Chapter 3: An explanation of the neurological roots of suffering
- Chapter 4: An explanation of the theory behind overcoming suffering
- Chapter 5: An outline of the core practices for overcoming suffering
- Chapter 6: An outline of more advanced practices for overcoming suffering
- Chapter 7: A call to pursue a life of hope

CHAPTER 2

The Problem With Suffering

As you will see throughout this book, suffering is a complex process—both biologically and psychologically—but that is not the problem with suffering. Actually, the problem with suffering is disarmingly simple:

We don't *need* to suffer.

Let me explain. We suffer for many reasons: unfortunate life circumstances, natural disasters, and intentional acts of hostility and aggression. Life is, undeniably, filled with painful experiences. We all know people with entirely valid reasons for their suffering, those whose lives are marked by loss, neglect, or trauma that would leave anyone hurting.

But the greatest problem I encounter in my life and in the lives of those I work with isn't the unavoidable challenges life throws at us. The real problem is intrinsic suffering: the suffering we generate within our own minds. You see, even when life's external events are dire, suffering is not an inevitable result—it is our internal reaction to those events that creates suffering. As we will explore in the next chapter, much of our suffering originates in the brain. It is perpetuated by where we place our attention, by the thoughts we nurture, and by the narratives we allow to define us. And because suffering is shaped internally, we have a choice.

We can consciously choose to suffer less.

The same underlying biological phenomenon—neuroplasticity—that enables suffering to clutch and then crush us in its ruthless talons can also help us to escape it. As suffering is a cognitive (and secondarily emotional) reaction to pain, and as our thoughts are subject to our volitional control, suffering is an option and not an inevitability. When we dwell less on suffering—ruminating less on past injury or fixating less on future fears—we malnourish those painful neural patterns, minimizing our suffering.

Sadly, this is a lesson most of us are never taught, largely because it's something our parents don't know.

Worse than this, many of us grow up in cultures that glorify suffering, whether through religious teachings or societal values that praise endurance and sacrifice. Think of the heroic suffering of the "Greatest Generation." They were born between 1901 and 1927, endured World War I as children, and suffered through World War

II in their primes. They have been championed for bearing their pain with intense stoicism and silence. But, while their resilience is admirable, it also perpetuated a belief that suffering is not only inevitable but noble.

This belief makes me sad, because it misses the larger truth that much of our suffering is unnecessary.

This is not to say that suffering can or should be eliminated entirely. And it doesn't take anything away from those who are forced to face and conquer major extrinsic disaster. Suffering and joy are two faces of the same coin, defined by their contrast. Without suffering, joy would lack meaning; without joy, suffering would lack hope. But while we cannot escape suffering entirely, we can choose to suffer less.

This is the power of conscious living: the recognition that we have agency over our internal experience, even when we cannot control our external circumstances. And so, as we start to explore the foundations of suffering, I hope that you're at least beginning to appreciate that, while happiness may be fleeting,

happi*er* is always possible.

When we embrace this perspective, the weight of suffering begins to lift because we recognize our power to respond in another way. Suffering is a choice. And when we choose differently, we begin to move toward the freedom and joy that are already within our reach.

CHAPTER 3

Understanding Suffering

This chapter lays the foundation for healing by helping you understand what suffering truly is: how it arises, how it embeds and perpetuates itself in the brain and body, and how it can be inherited. By distinguishing suffering from pain and examining its roots in trauma, narrative, and neural wiring, we begin to see that suffering is not inevitable. This insight prepares us to build a life of greater peace and opens the doorway to hope.

Suffering, at its core, is a deeply personal and subjective experience—a persistent mental and emotional reaction that extends beyond mere pain or discomfort. Unlike pain, which is a physical or sensory signal alerting us to immediate harm, suffering is an enduring state of inner turmoil. It is a profoundly emotional process, shaped by our thoughts, our memories, and our reactions to life's hardships.

Given its ubiquity, it is not surprising that both characters in our story experienced suffering.

Hope, a young girl with her whole life ahead, experienced an ache that she could not easily articulate yet felt keenly. While painful losses initiated Janet's journey of sorrow, her suffering became a longstanding companion, tied to regrets that burrowed into her psyche over time. Her suffering endured because of a repeating narrative that she held tightly; it was not just a reaction to past events but an identity, a lens through which she viewed herself and her life. Unlike the transient pain of her loss and abandonment, suffering became woven into her self-perception, influencing her thoughts and behaviors long after the initial injuries.

Janet and Hope's experiences reflect the reality that suffering often arises not from events themselves but from our internal reactions to them—it is a product of the mind.

Our capacity for suffering is paradoxically tied to the very faculties that allow us to love deeply, imagine vividly, and hope unreservedly. The same neural sophistication that grants us memory, empathy, and anticipation also renders us vulnerable to enduring emotional wounds, making suffering an intricate neurobiological process.

Throughout recorded history, stories of suffering have captured the essence of human resilience and vulnerability. Consider the biblical figure Job, who endures unimaginable trials and losses, his story illustrating the depths of despair and the search for meaning amid suffering. In the ancient Greek epic of Odysseus, we encounter a hero who, despite his courage and strength, endures years of separation, loss, and unfulfilled longing—a tale of suffering that stems not only from his external trials but also from his internal yearning for home and peace. In more recent history, we can look to figures like Anne Frank, whose diary chronicles the brutal realities of war and reveals her internal struggle to maintain hope and dignity in the face of profound suffering. These stories underscore that suffering, whether physical or emotional, transcends individual experience and speaks to a shared human condition.

Yet the very universality of suffering offers a surprising source of hope. If suffering is a natural reaction built into our neurobiology, it also means that we possess the means to understand and perhaps reshape it. This awareness invites us to see suffering not as an insurmountable burden but as a shared aspect of our humanity that can be approached with compassion and curiosity.

In this chapter, we will explore the ways in which our brains encode and perpetuate suffering. We'll examine the brain's role in interpreting pain and delve into the mechanisms by which our internal reactions and responses—rather than our external circumstances—often shape our suffering experience. Through this exploration, we will appreciate that suffering is not a fixed aspect of life but a dynamic process that can be understood, managed, and even transformed. In doing so, we follow in the footsteps of countless others who have confronted suffering with courage, resilience, and the hope of finding peace beyond the struggle.

The Suffering/Joy Continuum

To begin, let's take a moment to consider suffering from a philosophical perspective. If suffering represents a state of inner turmoil and prolonged emotional anguish, its opposite might best be described as inner peace and joy—a grounded state where we are free from the weight of unresolved pain and unburdened by lingering distress.

Unlike fleeting happiness or superficial pleasure, inner peace and joy reflect a lasting sense of well-being, one that endures through life's inevitable ups and downs. They do not create a life free from challenge but a state wherein challenges do not become entrenched. In this state, we meet adversity with calm acceptance and an open heart. Janet began to enjoy this new state as she was released from the iterative grip of past wounds—a shift from seeing herself as bound to suffering to understanding herself as capable of peace and joy.

The concept of duality—of irreconcilable, dialectical opposites—has been central to human philosophy, an unresolved conundrum. From light and dark to life and death to order and chaos, we see examples of forces that seem to oppose each other yet remain intimately connected, each giving the other its meaning and shape. Not only are they intrinsically bound, but they are inseparable; the one cannot exist in the absence of the other. You need a dark room to see the beam of a flashlight.

Suffering and joy represent one of these great dualities. They are often seen as two sides of the same coin or as two distant points on a continuum. In some philosophical traditions, this continuum is viewed as circular and hence infinite, suggesting that these states are interdependent rather than separate. As Carl Jung once observed, "There is no light without shadow and no psychic wholeness without imperfection."[1] The idea that peace and joy cannot exist without suffering suggests that, far from being irreconcilable, these states are part of an inextricably intertwined cycle, each revealing itself in contrast to the other.

We will explore suffering and inner peace not as isolated or fixed states but as markers on a continuum, a journey rather than a destination. This perspective opens the possibility that, although suffering may be an inevitable part of life, it need not be an insurmountable one.

As we navigate this continuum, we learn to see suffering not as an endpoint but as part of a transformative process, one that invites us into deeper awareness, resilience, and ultimately a state of enduring peace.

Consequences of Suffering

Before we dive into the causes of suffering, we must fully acknowledge its consequences. Why do we care about suffering in the first place?

Suffering, whether stemming from physical pain or emotional distress, exerts a massive toll on both the body and the mind. Its manifestations are multifaceted, affecting physiological health, mental well-being, and the quality of our relationships. Below are some of the many physical, mental, psychological, and social consequences of suffering.

Suffering, particularly when it becomes chronic, has a profound impact on the cardiovascular system.[2,3,4,5,6,7] Stress hormones like cortisol and adrenaline, released during periods of prolonged emotional distress, increase heart rate and blood pressure, placing a persistent strain on the heart and blood vessels. Over time, this can lead to hypertension, heart disease, and even stroke. Chronic stress significantly burdens the cardiovascular system, contributing to long-term risks for conditions like atherosclerosis and heart failure.

The gastrointestinal system is particularly sensitive to the effects of stress and emotional pain, often serving as a somatic outlet for suffering.[8,9,10,11,12,13,14] Chronic stress can exacerbate and possibly even cause conditions such as irritable bowel syndrome (IBS), peptic ulcers, and other digestive disorders, disrupting the gut's natural rhythms. When stress manifests physically through stomachaches, nausea, and other gastrointestinal issues, it further complicates an individual's ability to manage their suffering.

Suffering also compromises the body's immune system, leaving individuals more vulnerable to infections and illnesses.[15,16,17,18,19,20] Prolonged stress suppresses immune responses by altering the production of cytokines, proteins crucial for fighting off infections. This diminished immune function means that even minor ailments can persist longer and recur more frequently. Chronic stress reduces the body's resilience, making it harder to recover from illnesses and increasing susceptibility to new infections.

Emotional suffering often manifests as physical pain, particularly in the musculoskeletal system.[21,22,23,24,25,26,27,28] Stress-induced muscle tension is a common response to emotional distress, leading to tension headaches, migraines, and conditions like temporomandibular joint (TMJ) dysfunction. Persistent suffering can exacerbate these physical symptoms, creating chronic-pain conditions that further amplify the experience of emotional distress. Stress and physical pain often reinforce and escalate each other; this interplay creates a feedback loop that deepens both physical and emotional suffering.

Continuous suffering can lead to mood disorders such as depression and anxiety, characterized by persistent sadness, hopelessness, procrastination, and excessive worry.[29,30,31,32,33,34] Chronic stress is a significant driver of these conditions, disrupting emotional regulation and perpetuating negative thought patterns. These disorders not only magnify the experience of suffering but also diminish an individual's capacity to engage meaningfully with life.

Suffering takes a significant toll on cognitive function, impairing concentration, memory, and decision-making abilities.[35,36,37,38,39] Emotional distress consumes cognitive resources, leaving reduced capacity for tasks requiring focus and clarity. Chronic stress affects the prefrontal cortex, the area of the brain responsible for executive functions, exacerbating difficulties with attention and problem-solving. These cognitive impairments can hinder daily functioning, creating a sense of stagnation and helplessness that perpetuates suffering.

Sleep disturbances are a hallmark of prolonged suffering, as emotional distress often disrupts the body's natural sleep cycles.[40,41,42,43,44,45,46,47,48] Insomnia, restless sleep, and nightmares are common in individuals experiencing chronic stress and emotional pain. These disruptions create a viciously self-perpetuating cycle: Lack of sleep exacerbates emotional instability, while heightened emotional distress makes restful sleep more elusive.

Suffering often isolates people, creating barriers in their relationships.[49,50,51,52] Emotional pain can make it difficult to articulate feelings or to read others, leading to misunderstandings and conflicts with loved ones. Stress-induced irritability and anger strain communication and create distance between individuals. These barriers prevent meaningful connection, driving isolation and amplifying suffering.

Suffering can also diminish intimacy in romantic relationships.[53,54,55] Emotional distress and physical pain reduce sexual desire and closeness, weakening bonds between partners. Stress is a major factor in sexual dysfunction, further complicating relationships already strained by suffering. This loss of intimacy often intensifies feelings of disconnection and resentment, perpetuating cycles of emotional pain within partnerships.

Broader social withdrawal is a common response to prolonged suffering.[56,57,58,59,60] Individuals in pain may feel misunderstood or burdensome, leading them to retreat from loved ones and social interaction. This isolation deepens emotional distress, creating a self-reinforcing loop of loneliness and pain that cuts individuals off from the support systems that could otherwise help alleviate their suffering.

Psychological and emotional suffering can significantly impair decision-making, leading to adverse economic outcomes.[61,62,63,64] Individuals experiencing mental health challenges, such as depression or anxiety, often exhibit increased impulsivity and inconsistency in their financial choices, resulting in behaviors like compulsive spending or making poor investment decisions. Additionally, emotional distress can heighten vulnerability to financial exploitation, making individuals more susceptible to scams and fraud.

These physical, emotional, relational, and even financial consequences of suffering reveal its pervasive negative impact, reaching into every facet of a person's life. Fortunately, a deeper understanding of the neuroscience of suffering offers us a way to avoid these consequences.

Pain Speaks; Suffering Echoes

Understanding the distinction between pain, chronic pain, emotional pain, and suffering is essential before we explore the neuroscience of suffering. While these terms are often used interchangeably, they represent distinct phenomena with different origins, mechanisms, and impacts on the body and mind. By clarifying these concepts, we lay the groundwork for understanding how suffering arises and, most importantly, how it can be transformed.

Pain is a biological mechanism designed to protect us from harm. It is the body's alert system, signaling when something has gone wrong—an injury, illness, or threat to physical integrity. Pain can be categorized into two primary forms: acute pain and chronic pain. *Acute pain* is short-term pain that arises in response to a specific cause, such as a broken bone, a burn, or surgery. Acute pain typically resolves once the underlying issue is addressed and healing occurs.

Chronic pain is persistent pain that continues for weeks, months, or even years, often long after the original cause has healed or without a clear physical trigger. Chronic pain is associated with changes in the nervous system, such as *central sensitization*, where the brain and spinal cord become hyperreactive to pain signals. Conditions like arthritis, fibromyalgia, and neuropathy are common examples.

While pain originates in the body, it also engages the brain, where it is processed and interpreted. This becomes particularly relevant when we consider how the experience of pain can evolve into suffering.

Suffering extends far beyond the physical sensation of pain. It is a subjective and often prolonged emotional reaction to adversity, rooted in the mind's interpretation of, memory of, and reaction to experiences. It arises when we attach meaning, narratives, or unresolved emotions to painful or challenging events. Unlike pain, which has a clear physical origin, suffering can exist without any physical trigger at all.

Although chronic pain and suffering frequently overlap, they are distinct in their origins and mechanisms. Chronic pain is sustained by physiological changes in the nervous system, such as nerve damage and inflammation, whereas suffering is perpetuated by mental and emotional processes like rumination, unresolved trauma, and negative self-narratives. Chronic pain often requires medical or physical interventions. Suffering, by contrast, can be alleviated by shifting the mind's perspective, fostering awareness, and adopting practices that promote emotional well-being, as you'll see in Chapter 4.

Chronic pain and suffering often reinforce each other. Chronic pain can lead to suffering when it creates feelings of helplessness, frustration, or isolation. Similarly, suffering can amplify the perception of chronic pain, as emotional distress heightens the brain's sensitivity to physical discomfort. This interplay makes it important to address both aspects simultaneously, recognizing where physical interventions are needed and where the mind's perspective can be shifted.

Emotional pain occupies a unique and vital place in the broader discussion of suffering. Unlike physical pain, which arises from tangible injuries or conditions within the body, emotional pain emerges from deeply impactful nonphysical experiences such as the loss of a loved one, the end of a meaningful relationship, the sting of rejection or failure, or the sorrow of unmet expectations and unfulfilled potential. While it lacks a physiological origin, emotional pain is deeply felt, often with an intensity that equals or even surpasses that of physical pain.

In many ways, emotional pain serves as a bridge between the realms of pain and suffering. It shares similarities with physical pain in terms of how the brain processes it, engaging the same neural pathways and regions. This neurological overlap explains why emotional wounds, such as heartbreak or grief, can feel so visceral. Yet emotional pain does more than mimic physical pain—it also plays a significant role in sustaining suffering. The death of a loved one, for example, might initially trigger acute emotional pain, but it is rumination and unresolved sorrow that transform it into prolonged suffering. While Janet's pain originated from real psychological injuries, her suffering was sustained by the emotion-laden narrative she carried with her every day.

Understanding emotional pain as the easy-to-overlook link between pain and suffering provides critical insight into its role in our lives. By addressing emotional pain directly—examining its origins, its neurological underpinnings, and its connection to our inner narratives—we open the door to transforming it.

Emotional pain offers us a point of intervention,

where we can begin to untangle the threads of suffering and move toward a life of greater inner peace and joy.

But before we deepen our understanding of how pain becomes suffering, we need to understand a little more about the biology of pain and, in particular, the role of our magnificent brain in the experience of pain.

The Brain's Role in Pain

The brain's primary function is survival, and so it is wired to detect and respond to threats. Pain is a critical signal, a protective mechanism designed to alert us to potential harm and prompt immediate action. The full process of pain begins with an external stimulus, such as an injury, which activates sensory neurons called *nociceptors* (pain receptors) that are widely distributed throughout the body—in skin, muscles, joints, bones, internal organs, and even the outer lining of the brain. These specialized neurons send electrical signals along *afferent nerves*, which carry information about the painful stimulus to the spinal cord and then into the brain.

Once in the brain, these signals are processed in the *somatosensory cortex*, which localizes the pain, and in the *anterior cingulate cortex*, which evaluates both its physical and its emotional intensity.[65,66] Special regions of the brain then focus the brain's attention on the incoming pain messages, while amplifying them and enabling survival-based decision-making.

Regulatory structures such as the *thalamus* act as relay stations, helping to process and distribute pain signals to other parts of the brain while further defining the pain and triggering aversive emotions in reaction to it.[67] The *limbic system* plays a central role in this amplification process.[68,69,70,71,72] The *amygdala* assesses the threat and initiates the body's stress response by assigning emotional significance to the incoming sensory information. The amygdala thus heightens the experience of pain through the addition of fear or anxiety.[73,74] The incoming message is then distributed to the *hippocampus* to store the painful memory, making it readily accessible for recall.[75,76,77,78] Finally, the brain initiates an *efferent response*, sending signals back through motor pathways to influence behavioral and physiological reactions to pain, such as withdrawing a hand from a hot surface.

Emotional pain follows a remarkably similar pathway.[79,80] The anterior cingulate cortex, for example, lights up during both physical injuries and emotional wounds. This overlap explains why emotional pain feels so immediate and real—it engages the brain's pain-processing systems even in the absence of a physical injury.

The brain's ability to focus on and amplify pain—whether physical or emotional—is a double-edged sword. While it serves an evolutionary purpose by ensuring that we avoid ongoing or repeated harm, you can see how it also shifts the pain experience into a suffering experience.

Emotional memories are not static; they are easily triggered by subtle cues, such as a smell, a song, or even an unrelated stressful event. When reactivated, these memories evoke a fresh wave of emotional pain, as if the original event was happening again, creating a cycle of recurrent suffering.[81,82,83] The amygdala often remains hyperactive, exaggerating the brain's perception of pain even in the absence of a new injury.[84,85,86,87,88] The *prefrontal cortex*, responsible for rational thinking and emotional regulation, struggles to counterbalance this hyperactivity. Research shows that emotional regulation—the brain's ability to manage and moderate emotional responses—is critical to preventing emotional pain from escalating into prolonged suffering. However, when the prefrontal cortex is overwhelmed or underactive, the amygdala takes over, driving the heightened emotional responses that can spiral into chronic suffering.[89,90,91,92]

While pain begins as a transient experience, emotion primes and sustains it.

Emotions such as fear, shame, and anger are particularly potent in this regard, as they amplify the brain's stress response, making pain feel more intense and harder to release, often with deleterious long-term effects.[93,94]

The science behind this transformation lies in the brain's *neuroplasticity*—its ability to rewire itself based on repeated experiences.[95] Although better known in contemporary science for its ability to positively reshape our lives, this same plasticity can work against us, reinforcing patterns of suffering as our brains repeatedly revisit painful memories or anxious anticipations. This amplification is further fueled by rumination, a cognitive process in which the mind revisits and dwells on painful experiences, strongly reinforcing the neural pathways associated with suffering.[96,97,98,99,100,101] Over time, this can lead to a state where the brain becomes hypersensitive to pain signals, both physical and emotional, creating a self-perpetuating cycle of suffering.[102,103,104]

The Evolutionary Bias Toward Pain

Another facet of our protective design sets us up to suffer: In its effort to keep us alive at all costs, the human brain has developed a remarkable ability to focus on negative experiences. This evolutionary adaptation—known as the *negativity bias*—ensures that we remain vigilant to potential threats, cataloging them deeply in our memory so that we can avoid danger in the future. This once-protective mechanism becomes dysfunctional in modern life, however, where physical threats are less common and the challenges we face are more psychological.

The negativity bias means that our minds are not impartial observers; they prioritize negative experiences and hold onto them more tightly than positive ones. You may remember that negative memories are stored in the hippocampus, which works in conjunction with the amygdala to store negative experiences more vividly.[105,106] Their storage triggers the release of stress hormones such as cortisol, which strengthen the neural connections responsible for encoding the memory.[107] These negative memories are ready to be retrieved at a moment's notice should a similar danger arise. By contrast, positive experiences—which do not pose a threat to our survival—are often processed with less intensity and stored more fleetingly. This is why negative events stick to us, while positive ones often slide away unnoticed. Janet's memories of abandonment as a young girl, though decades old, remained vivid and emotionally charged, overshadowing the many moments of connection and kindness she likely experienced throughout her life.

This imbalance is not a failing of character but a feature of how the brain processes and encodes emotional experiences. Still, this asymmetry has major consequences for our mental and emotional landscapes.

Over time, the accumulation of negative memories forms the foundation of self-perpetuating cycles of suffering.

The brain, conditioned by past pain, begins to anticipate future harm, and our internal narratives become dominated by what has gone wrong rather than what has gone right. Janet's repeated fixation on her early losses is a vivid example of how negative memories, left unchecked, can shape our self-perception and create a persistent sense of sorrow. The stories we tell ourselves about our pain become the scaffolding for prolonged suffering, reinforcing the idea that the world is harsh and our lives are irreparably damaged.

Yet, while the negativity bias is deeply ingrained, it is not immutable. The human brain is astonishingly plastic, capable of rewiring itself in response to new experiences and intentional effort. In the next chapter, we will explore how the gift of neuroplasticity means that the same mechanisms that amplify negative memories can be harnessed to strengthen positive ones.

By now, I hope that you're beginning to appreciate that the suffering brain is not defective.

Janet didn't plan to suffer. Her brain, acting in her best interest, triggered a normal cascade of pain amplification and memory preservation. But, in its vigorous survival-oriented commitment, it also established a deeply embedded and enduring sorrow. Her suffering became entrenched, not just because of the intensity of her historic pain but because of how her brain continued to preserve, amplify, and focus on it in the present, with overwhelming consequences.

So, if you're reading this book because you feel you suffer more than is reasonable, I want you to know that your brain is not broken. Far from it. I hope that you now appreciate that your perfectly designed brain is just doing its job well. I hope that this chapter has also awakened some cautious optimism in you. After all, if we understand its underlying biology, shouldn't we be able to intervene and direct your magnificent brain away from suffering?

Before we get there, it is worthwhile to further explore the origins of suffering. How does it arise, within your brain and in your life?

The Impact of Childhood Trauma

Traumatic events experienced during childhood establish a rich breeding ground for suffering, shaping the developing brain in lasting ways.

Childhood is a period of extraordinary neuroplasticity, during which the brain is rapidly forming connections and adapting to its environment. This high plasticity makes children uniquely capable of learning and growth, but it also leaves them deeply vulnerable to the effects of trauma. Experiences of abandonment, neglect, or abuse during these formative years do not merely leave scars—they actively shape the wiring of the brain, creating patterns of emotional reactivity that can persist into adulthood.[108,109,110,111,112,113,114]

The stress caused by trauma, particularly when chronic, activates the brain's threat-detection system. This repeated activation creates a hyperreactive stress response, essentially priming the brain to see danger even where none exists. The prefrontal cortex often struggles to develop optimally under such conditions, leaving individuals with fewer resources to manage their emotional pain. Over time, this imbalance can lead to chronic anxiety, depression, and a heightened sense of vulnerability—the hallmarks of prolonged suffering.

Janet's mother's death and resultant abandonment of her left her with a profound sense of loss, one that became the foundation of her internal narrative. As Janet grew, this fearful abandonment narrative shaped her relationships and self-perception, reinforcing her feelings of unworthiness and isolation. The hyperreactive stress response formed in her early years meant that even minor triggers—an unreturned call, a perceived slight or insult—could awaken her deep-seated fears and send her spiraling into emotional pain. This nurtured a pervasive neediness that further perpetuated her isolation and suffering.

It is important to recognize that childhood trauma does not arise from only overtly catastrophic events. The process of growing into competence—of being stretched from a state of dependence at birth to functional autonomy in adulthood—inevitably induces stress and, at times, distress. While manageable challenges are essential for growth, excessive stress or unaddressed emotional pain can be problematic. The magnitude of this "stretch" determines the intensity of its impact: A small stretch creates stress. A larger stretch invokes distress. Anything larger than this is traumatic for the young and vulnerable human. When the stress of childhood becomes overwhelming, it lays the groundwork for the painful memories, fixed narratives, and suffering cycles that we explore in this chapter.

As a result of our inevitable ineptitude at birth, every single adult has been exposed to a range of traumatic experiences through their early lives. These childhood experiences, etched into the brain during its most formative years, can leave us trapped in a reactive state, where old wounds are easily reopened by present-day triggers. This is where most adult suffering originates.

But it isn't the only source of suffering.

Inherited Shadows

Much of our suffering is shaped long before we are conscious of it, but it's not all the consequence of childhood trauma. We can also inherit a suffering disposition. These "inherited shadows" come to us through a complex interplay of genetics and modeled behavior, embedding themselves into the narratives that shape our sense of self and our relationship with the world.

Recent research has illuminated how trauma and stress can imprint themselves on our biology, altering the expression of genes in ways that influence not only our health but also the emotional and psychological patterns passed down to our descendants. This intergenerational transmission of suffering helps explain why the shadows of trauma—such as fear, anxiety, and hypervigilance—often persist within families long after the initial events have faded into history. These are patterns of suffering that we inherit from our parents.

Epigenetics, the study of how gene expression is influenced by environmental factors, has revealed that trauma can leave identifiable marks on our DNA without altering the genetic code itself. These changes, often in the form of a chemical alteration to chromosomal proteins (known as *DNA methylation*), effectively switch certain genes on or off, creating biological responses that persist across generations (sometimes even skipping generations).[115,116,117,118] For example, studies on the descendants of Holocaust survivors[119,120] have demonstrated epigenetic changes to a gene involved the regulation of cortisol—a hormone critical for managing stress. These epigenetic marks, likely a response to the extreme trauma endured by the survivors, have been observed in their children, who exhibit heightened stress reactivity even without direct exposure to similar events.

Further evidence comes from the Dutch Hunger Winter,[121,122,123,124] a period of famine during World War II. Children who were in utero during the famine were found to carry specific epigenetic signatures that affected their health throughout their lives. Studies on the sons and grandsons of Civil War prisoners of war[125,126,127] found they had higher mortality rates compared to their peers, suggesting that the extreme stress endured by their fathers and grandfathers had epigenetic effects carried on the Y chromosome. Such findings offer a biological explanation for the enduring shadows of historical trauma, providing context for how major societal upheavals can reverberate across generations.

I want to emphasize that, while epigenetics provides a powerful lens for understanding the intergenerational transmission of trauma, it is not deterministic. The interplay between genetic predisposition, epigenetic modifications, and environmental influences is tremendously complex. While these inherited patterns may shape an individual's vulnerabilities, they do not predetermine their fate.

This understanding offers a crucial layer of hope: Even in families burdened by generations of suffering, the cycle can be interrupted through awareness, healing, and intentional action. Just as trauma can leave its genetic mark, so healing and growth can create new imprints.

But the inheritance of suffering is not limited to genetically coded or epigenetically transferred biology. It also manifests in the behaviors, emotions, and patterns modeled by caregivers and absorbed by young children.

The stories we inherit from our families can reinforce this generational transmission of suffering. These stories—of struggle, loss, or fear—become part of the narratives that define a family's identity. For a child, these inherited narratives often create a lens through which they view the world, shaping their expectations, their fears, and even their sense of self-worth. In computational language, when we're young and our brains are developing, we copy and paste our parents' attitudinal and behavioral algorithms and recurring familial patterns. As we grow, these shadows settle into our subconscious, influencing our thoughts, emotions, and reactions in ways we may not fully recognize or understand.

Reflected Suffering

The final origin of suffering I'd like to discuss may surprise you, as it is the consequence of a trait that is commonly (and rightly) held to be advantageous: empathy.

Mirror neurons, first identified in the early 1990s, are specialized brain cells that "mirror" the actions and behaviors of others.[128] This mirroring mechanism enables individuals to understand and imitate others, a faculty that plays an active role in cognition, language, and empathy.

In the context of suffering, these mirror neurons play a pivotal role in our ability to empathize with others' pain. When we witness someone experiencing discomfort, our mirror neuron system simulates the observed experience. We literally feel their pain and experience their suffering. This is particularly true for infants and children, who are exquisitely attuned to their parents' emotional states, unconsciously mirroring and internalizing them as part of their own emotional framework.

This empathic resonance, while a powerfully prosocial behavior, can have complex and undesirable implications for our own well-being. By internalizing another's pain, we may inadvertently trigger or amplify our own suffering, especially if the observed distress resonates with our past personal experiences. In the absence of awareness or emotional regulation, this vicarious suffering can become deeply ingrained, reinforcing existing patterns of distress and even reactivating personal wounds from our own experiences.

This process is particularly insidious when suffering is normalized within familial or social structures, where the repeated exposure to another's distress strengthens neural pathways that sustain suffering across generations. Over time, this mirrored suffering can manifest as chronic emotional pain, shaping our perceptions, behaviors, and interactions in ways that make suffering feel inevitable rather than optional.

In this way, suffering becomes a legacy, passed from one generation to the next, not only through words but through unspoken emotional undercurrents and nonverbal signaling.

The Embodiment of Suffering: The Fixed Narrative

As we have seen, suffering often begins with a singular event—a loss, a rejection, or a mistake. But the mind has a way of compounding that initial pain, layering it with self-inflicted wounds that deepen and prolong the experience of suffering. The mind solidifies the suffering into a story that shapes how we perceive and explain ourselves, our relationships, and the world around us.

Janet believed that she had destroyed her marriage. In that single thought lay the weight of her shame, self-blame, and regret. This story became more than a reflection of her past—it became the foundation of her identity and the scaffolding for her continued suffering.

The stories we tell ourselves arise from the brain's natural processes of memory and thought. When we revisit them—repeating the thoughts, emotions, and meanings tied to those experiences—the brain reinforces the underlying neural pathways that support them, making them more automatic, quicker to activate, and harder to disrupt. Over time, this repetition organizes the memories into a *fixed narrative*: a studiously reinforced mental script that defines not only what happened but also what it means about who we are.

Two of the most insidious voices in these narratives are *blame* and *shame*, which not only exacerbate the original wound but also create their own self-perpetuating cycles of distress. These emotions, tied to our sense of self-worth and responsibility, can turn a fleeting moment of pain into an enduring state of suffering.

Blame often emerges as the mind searches for a reason for the pain, a target on which to pin responsibility. In many cases, this blame turns inward. Children often blame themselves for bad things that happen. We see it commonly in cases of divorce where a child blames themself for their parents' estrangement. Janet hosted such a belief deep within her subconscious. Later in her life, she repeatedly told herself the story of how her neediness and emotional clinging contributed to the demise of her marriage. This narrative not only reinforced her feelings of unworthiness but also kept her trapped in a loop of self-recrimination.

Similarly, shame—closely related to blame—magnifies suffering by making the pain deeply personal. While blame seeks a target for responsibility, shame attacks the self directly, embedding the belief that *I am flawed* or *I am not enough*. Janet's shame over her perceived failures in relationships added an additional layer of suffering, distancing her not only from others but also from herself.

One of the most significant drivers of this embedded narrative process is rumination, which creates a self-reinforcing loop: Each time the mind revisits a painful memory or negative thought, it reactivates the associated emotions and strengthens the neural pathways tied to that pain. These repeated iterations not only trap the suffering victim in their suffering but also amplify its emotional impact.

Janet's repeated narrative of "I destroyed my marriage" shows how rumination fuels self-inflicted suffering. With each repetition, her brain reinforced the feelings of blame, shame, regret, and unworthiness tied to that story, making it harder for her to break free from the cycle. Janet's story shaped her perception of relationships and her role within them. It influenced how she engaged with—or withdrew from—others, how she interpreted moments of connection, and how she made choices about relational intimacy.

What makes these narratives so insidious is how they percolate in every corner of our lives. Like an invisible hand steering the wheel, these narratives shape our lives without our awareness, perpetuating the suffering they represent. This is the dangerous power of fixed narratives: They not only describe our suffering; they sustain it, initiating and maintaining behaviors that deepen the very pain they articulate.

When we are unaware of these narratives, they continue unchecked, becoming so entrenched that they feel inseparable from who we are. Yet, while these "fixed" narratives feel deeply uncontrollable, they are not. The same mechanisms that create them also offer the potential for change. The brain's neuroplasticity allows for the revision of thought patterns, through the weakening of old neural pathways and the creation of new ones.

Recognizing the stories we tell ourselves is the first step.

Once we see the narrative for what it is—a construct of the mind rather than an absolute truth—we can begin to untangle its hold on us.

The Foundation for Hope

Our journey together through the nature of suffering has revealed its deeply embedded biological and psychological mechanisms. We have explored the brain's deliberate processes for storing, amplifying, and projecting painful experiences that have happened to us.

Suffering, as we have seen, is not simply the pain of external events but the brain's internal reaction to those events—a process shaped by evolution, modeling, memory, and habit. It begins with the brain's survival-driven negativity bias, which prioritizes painful experiences for storage and recall, and extends into the active subconscious as these experiences form recalcitrant narratives that guide our thoughts, emotions, and behaviors.

We have examined how suffering becomes a self-perpetuating cycle through the strengthening of neural pathways and the brain's tendency toward rumination. These processes create a default survival-oriented neurobiological framework wherein suffering becomes less about the original event and more about the mind's repeated engagement with it.

The brain, through its plasticity, adapts to what it experiences most frequently—whether that is curiosity, joy, and peace or fear, pain, and despair.

This scientific understanding underscores why suffering feels so unyielding and why escape from it often seems elusive.

This chapter has laid the groundwork for understanding suffering as both an adaptive process and a mechanism that can spiral into dysfunction. Armed with this understanding, we are better equipped to approach the next steps: learning how to consciously engage with these processes to rewrite the narratives, disrupt the cycles, and take control of our internal reactions. The challenge is not to eliminate suffering entirely—it is a fundamental part of life—but to learn how to navigate it skillfully, reducing its impact and enhancing our capacity to move toward a state of inner joy and peace.

CHAPTER 4

Awakening Hope

On our journey together, this chapter marks the turning point from understanding suffering to beginning its transformation. You'll be introduced to foundational shifts in mindset and practical tools that help awaken hope. Drawing on both body and mind science, these early interventions remind us that healing is possible, that agency puts control into our hands, and that we are empowered to effect meaningful change—starting today.

The journey through suffering is a deeply personal one, yet it follows patterns that are universally human. In the previous chapter, we explored suffering's origins in both the mind and the body, uncovering the mechanisms that entrench it and the ways in which we unwittingly keep ourselves trapped in cycles of pain. We also confronted the uncomfortable truth that much of our suffering—though deeply felt—is unnecessary.

Now we turn to the most important question: *How do we transform suffering into inner peace, resilience, and joy?*

The remaining chapters are about solutions. They are about moving beyond the intellectual understanding of suffering and stepping into practical, actionable change. We will explore how the mind can be trained, the brain rewired, and suffering reframed—not through wishful thinking, but through systematic, science-backed strategies that align with our deepest potential for healing and growth.

There are moments in life that shift our understanding of suffering, epiphanies that offer us glimpses of hope and the power of choice. One such moment came to me early in my career as a junior doctor. I cared for a young man who had been paralyzed by a gunshot wound to his spinal cord. He lay in his hospital bed unable to move, his body ravaged by necrotizing bedsores. His physical suffering was immense, but what struck me even more deeply was his emotional despair. He had lost not just his mobility but, in his mind, his entire future.

In the face of such devastation, I felt powerless to ease his suffering. It was his sister who showed me what true healing looks like. I trained her in the care of his wounds, and with extraordinary love and patience, she nursed him back to health. Over time, his sores healed, his body stabilized, and his spirit was restored. Although dependent on his sister's ongoing physical support, he returned to the hospital regularly—not as a patient but as a beacon of inspiration, sharing his story with staff and patients, radiating gratitude and resilience.

That experience transformed my understanding of suffering. It was not the young man's paralysis that defined his suffering but the meaning he assigned to it. His sister's unwavering care changed his outlook, allowing him to reclaim his life in ways that went beyond physical recovery. This was a profound revelation: Suffering is not just an experience; it is a choice.

At the heart of this book is the recognition that, while pain is often unavoidable, suffering is malleable. It can be shaped by our perception, our mindset, and the actions we take in response to adversity. We are not powerless. The mind is a tool that can be trained, and just as suffering can become ingrained, so too can resilience, peace, and joy.

As we move forward, we will explore the practical steps to awaken hope and reclaim authority over the mind. We will examine how the brain's neuroplasticity offers a road map for healing. We will discover how to reframe suffering, harness resilience, and break free from inherited cycles of pain. Most importantly, we will see that the power to transform suffering does not lie in external circumstances but in the choices we make, moment by moment.

The First Steps Toward Transformation

Transforming suffering is not an abstract concept—it is an intentional process, a deliberate act of self-liberation. Such change requires both introspection and practical action. It is about finding the balance between acknowledging our pain and actively seeking ways to heal. Insights from neuroscience provide invaluable tools for this process, helping us understand how resilience, positive thinking, and emotional regulation can rewire our brains for healing.

True transformation beyond suffering begins with two fundamental shifts in mindset. They mark the transition from being a passive participant in pain to playing an active role in making pain a force for healing.

First, become comfortable in discomfort. Paradoxically, the more we resist suffering, the more power it holds over us. Suffering does not break us; it is our resistance to it that makes it feel unbearable. So, the next time discomfort arises, remind yourself, *This is difficult, but I can bear it.* If you hear yourself saying, "I cannot tolerate the pain," you will make it so. So, don't say it! Instead, pause. Observe the discomfort without judgment, neither fighting it nor amplifying it. Then say (out loud if necessary), "This pain is bearable." Simply acknowledging that suffering is survivable begins to reduce its intensity, weakening its hold over the mind.

Second, find value in suffering. We must embrace suffering not as an adversary but as an inevitable and even valuable part of life. When we reframe suffering as an opportunity for growth rather than a threat, it becomes a teacher, offering lessons that deepen our self-awareness and resilience. In difficult moments, ask yourself, *What is this teaching me?* By shifting from a mindset of resistance (*Why is this happening to me?*) to one of curiosity (*How can I grow from this?*), we begin to reclaim authority over our suffering. The brain follows the narratives we reinforce, and choosing a perspective of growth rather than despair reshapes the way we experience pain.

Suffering is an undeniable part of the human experience, but it is also a catalyst for growth. Every challenge presents an opportunity to not just endure but evolve. The power to transform suffering begins with shifting from reactivity—where suffering dictates our actions—to thoughtful awareness, where we pause, assess, and respond with intention. Eventually, this shift leads to proactive resilience, where suffering no longer controls us and we consciously build the mental and emotional frameworks that allow us to thrive. By practicing these two steps together—embracing discomfort as bearable and seeing it as an opportunity for growth—we transform suffering from an oppressive force into a source of strength and wisdom.

Initially trapped in a cycle of self-inflicted pain, Janet clung to her past narratives, unknowingly reinforcing her suffering. Yet through Hope's guidance, she began to see her suffering not as an inescapable fate but as a process she could influence. Janet's story is a metaphor for the path available to each of us: to shift from suffering to empowerment and from despair to hope.

In the coming chapters, we will explore how neuroscience supports this transformation, offering systematic strategies for building resilience, reframing pain, and reprogramming the mind for healing. But before we turn to the elegant mechanics of your brilliant brain, let's explore two simple, yet powerful, approaches to overcoming trauma.

Talk the Talk: Revisiting Trauma

Suffering is often deeply intertwined with past traumas, whether recognized or buried beneath layers of unconscious defense mechanisms. Trauma, particularly when left unprocessed, solidifies into persistent suffering, manifesting as emotional distress, intrusive thoughts, or even physical symptoms. You see, the brain does not differentiate between past and present when it recalls traumatic events—this is why reliving trauma can feel as visceral as the original experience. However, this same mechanism offers a powerful avenue for healing. By consciously revisiting painful memories in a structured way, we can begin to reframe our relationship with them.

Janet's childhood experiences of abandonment were not just memories but active forces shaping her present emotions and behaviors. Yet, through her dialogue with Hope, she began the process of reeducation by revisiting significant memories, each one holding a lesson.

In doing so, Janet engaged in emotional regulation practices. With Hope as her guide, Janet began to reflect on her emotions and to engage in self-inquiry—taking her first steps toward healing. She gradually shifted from reactive suffering to conscious processing, reinforcing the truth that emotional mastery is a learnable skill. Her journey highlights that emotional regulation is not about avoiding emotions but about learning to navigate them with awareness, self-compassion, and intentional response. These are powerful modalities, all accessed through the simple act of talking.

At the same time, Janet employed a valuable healing tool: *storytelling*. Simply telling her story helped Janet process her suffering. You will learn later how this simple technique invites cognitive restructuring, allowing individuals to reframe their pain in new, empowering narratives.

Without realizing it, Hope invoked an equally powerful tool: *compassionate listening*. One of the most effective techniques in processing trauma is the presence of a *compassionate witness*, someone who can reflect back understanding, validation, and hope. Trauma often isolates us, creating the false belief that we are alone in our suffering. Hope, with her gentle and inquisitive nature, provided Janet with a safe space where she could verbalize her deepest fears without judgment. The relationship was similarly beneficial for Hope; listening with compassion and patience has been shown to reduce stress and promote healing through oxytocin release. Supportive relationships—whether with trusted friends, mentors, or therapists—help reframe pain, transforming it from an oppressive weight into a shared burden that can be carried more lightly.

Finally, together, Hope and Janet deployed another well-described scientific phenomenon: *healing through connection*. Building and maintaining strong social connections can provide emotional support and a sense of belonging, which are crucial for mitigating feelings of isolation and suffering. Both compassion for others and self-compassion activate the parasympathetic nervous system, which induces the physiological state of rest and recuperation. This fosters a sense of safety and peace.

You can employ similar practices in your own life to begin transforming your suffering into strength.

Try guided visualization. Picture the past traumatic event, but reimagine it from a place of safety. See yourself as the observer, not the victim, and consciously rewrite the emotional tone of the memory.

Experiment with journaling. Writing about traumatic events can activate the prefrontal cortex, which helps regulate emotional responses, making it easier to process and release stored pain.

A Word of Caution: While revisiting trauma can be deeply healing, it can also be distressing. If memories feel overwhelming or trigger severe distress, it is essential to seek the guidance of a trained professional. Trauma-informed therapy provides the necessary structure and support to navigate painful recollections safely. If there is any doubt about whether to embark on this journey alone, err on the side of caution and seek professional support.

Walk the Talk: The Body Heals the Mind

The mind and body are inextricably linked, and unprocessed trauma often finds a home within the body itself. As Bessel van der Kolk explores in *The Body Keeps the Score*,[129] traumatic experiences leave imprints not just in our memory but in our muscles, nervous system, and physiological reactions. Healing trauma is not just about rewriting mental narratives; it is about erasing the score that trauma has kept in the body, allowing us to reset and continue playing the game of life with greater ease and resilience. The simple measures mentioned below all direct the body toward restoration. By integrating mental and physical approaches, we create a holistic pathway toward healing, freeing ourselves from suffering not just in thought but in our very physiology.

Exercise regularly. Regular physical activity releases endorphins, the body's natural painkillers and mood elevators. Exercise also lowers cortisol and adrenaline levels, helping to break the cycle of chronic stress.

Try yoga and tai chi. These mindful movement practices help regulate the nervous system, encouraging a shift from a hyperaroused (fight-or-flight) state to a relaxed (rest-and-digest) state. Studies show that trauma-sensitive yoga can reduce PTSD symptoms and restore a sense of bodily autonomy.

Focus on breathwork. Deep breathing exercises (such as diaphragmatic breathing) help regulate the vagus nerve, calming the body's stress response.

Get chilly with cold therapy. Exposure to cold (such as cold showers or ice baths) has been shown to reduce inflammation, lower stress hormones, and create a sense of resilience in the nervous system.

Reclaim the body's narrative. Trauma often robs us of a sense of ownership over our body. Practices that reinforce physical presence—such as massage therapy, dance, or simple grounding exercises like walking barefoot on grass—can help reestablish a connection between mind and body.

Janet and Hope intuitively combined these two early lessons: talking the talk and walking the walk. In their compassion and companionship, and while exercising their minds and bodies together, they discovered that together they could face any darkness with courage. So, if you don't read any further in this book, then perhaps the most elementary takeaway will be that, to address suffering, you should walk and talk with a good friend.

It's now time to start the serious work of liberating you from habitual, unnecessary suffering. It's time to start working directly with your magnificent brain.

Nature's Miracle: The Brain's Capacity for Healing

As we discussed in Chapter 3, the human brain—once thought to be a rigid and fixed structure—is now understood to be astonishingly adaptable. This neuroplasticity is the fundamental scientific reason for hope. No matter how ingrained our patterns of thought and emotion may seem, no matter how long we have suffered, the truth is that we are not prisoners of our past. The neural pathways that sustain suffering—the deeply embedded habits of negative thinking, fear, and self-criticism—can be restructured, replaced by pathways that support resilience, inner peace, and joy.

One of the most fundamental principles of neuroplasticity is that repetitive activation of a neural pathway strengthens that pathway, reinforcing the sequence of thoughts, emotions, and reactions that follows. With time, the brain defaults to these well-established routes, channeling new thoughts and emotions down the same familiar pathways rather than considering alternative approaches. This explains why suffering can feel so automatic and why, even in the absence of immediate triggers, we find ourselves caught in negative thought loops.

Yet, just as the mind solidifies pathways of suffering, it also retains the ability to forge new paths, rerouting traffic in more constructive, healing directions. Research in cognitive-behavioral therapy (CBT) and mindfulness-based stress reduction (MBSR) has shown that individuals who consistently engage in practices that shift their focus toward positive or neutral experiences exhibit measurable changes in brain structure, particularly in areas associated with emotional regulation and resilience.

The same biological process that entrenches suffering can, with intentional effort, be redirected toward inner peace and joy.

For years, Janet's brain reinforced a painful narrative of abandonment and unworthiness. Each time she recalled past wounds, she unknowingly strengthened those neural pathways, deepening her suffering. Every recollection and every self-critical thought further entrenched the well-worn grooves of her distress, ensuring that even neutral situations could trigger familiar cycles of pain.

Yet through her interactions with Hope, she introduced new perspectives and experiences—ones rooted in self-compassion and understanding. Each time she chose to interpret a situation differently, to challenge an old belief, or to pause before reacting, she was taking another tentative step onto an alternate path—one that, with time and repetition, could become the new default. Over time, these small choices accumulated, allowing her to gradually release the weight of old pain and step into a new way of being.

The understanding of neuroplasticity liberates us from the notion that we are trapped by our past experiences or predispositions. By making intentional choices—in where we place our attention, in how we respond to triggers, and in the habits we cultivate—we can actively reshape the neural highways of our minds and steer our internal experiences in a new direction. The power to change is not theoretical; it is biological. And with this knowledge comes immense hope—because no matter how deeply suffering has been wired into our brains, transformation is always possible.

You Are Not Your Brain

The brain is, arguably, the most complex phenomenon modern science has ever attempted to understand. And paradoxically, the more we study it, the deeper its complexity reveals itself to be. Every new discovery opens doors to even greater intricacies—sophisticated networks of neurons, biochemical feedback loops, and dynamic interactions between cognition, emotion, fear, and behavior.

And it is precisely this complexity that keeps so many of us trapped in suffering. The sheer number of influences, pathways, and interwoven systems makes it nearly impossible to simply "think" our way out of suffering. The human mind is not a straightforward machine that can be manually overridden when it malfunctions. Instead, it operates on a deeply embedded set of algorithms—some of which serve us, and others of which perpetuate cycles of distress. To break free, we need a framework that simplifies the brain's vast and intricate workings into something we can manage.

For this reason, I have developed a model I call the *Brain Operating System (BOS)*. This framework does not attempt to capture the full, staggering complexity of the brain; that would be impossible. Instead, it provides a structured, accessible way to understand how the brain functions so that we can consciously influence it.

But before we can truly grasp the power of the BOS, there is one foundational truth that must be understood and fully adopted: We are not our brain.

Or, more precisely...

You are not your brain.

While the brain is the biological seat of perception, thought, and emotion, it is not the essence of who we are. I cannot emphasize this enough. Understanding this distinction is essential to gaining agency over our internal experiences and transforming suffering into joy.

One of the strongest indicators that we are not our brain is our ability to observe our own thoughts from a distance. If we were truly one with our brain, we would be completely absorbed by every thought and feeling that arose. We would be unable to take a step back and analyze its activity. However, we regularly engage in introspection, reflecting on our thoughts, emotions, and reactions as though from an outside perspective. This ability to witness our own mental processes proves that we are something more than just the brain's automatic reactions. When Janet began to recognize her narrative of abandonment and self-blame as something separate from her core identity, she took the first step toward change.

We often believe that, because we think something, it must be true. However, neuroscience and psychological research show that thoughts are merely the outputs of neural processes shaped by past experiences, emotions, and external influences. When we step back, we begin to see thoughts as transient mental events rather than absolute truths. From a distance, we can question our habitual thinking patterns, as Janet did when she challenged the automatic belief that she was unworthy of love. By seeing it as just a thought rather than an absolute reality, she began to weaken its hold on her.

When we are young, we begin to notice words and sentences forming in our minds—voices speaking, seemingly from nowhere. We look around and, seeing no one else in the room, naturally assume that we are the ones talking. In reality, it is our brain "thinking." When we reach adulthood, we intellectually understand that our brain cannot talk, but by then, the perceptual error has been deeply ingrained through years of repetition. We mistake the brain's thoughts for our own identity—we think we are our brain and its thoughts, feelings, and fears.

By de-identifying from our thoughts—saying, for example, "I am *having the thought* that I am unworthy" instead of "I *am* unworthy"—we create space for change. Janet's breakthrough moments hinged on this realization. When she shifted from an internal narrative that she was broken to an understanding that she was simply experiencing emotions based on this erroneous conclusion, she loosened suffering's grip and reclaimed control over her narrative.

Not all our thoughts originate from personal experience. Many are inherited, conditioned, or biologically programmed for survival. Recognizing this fact allows us to depersonalize distressing thoughts and emotions. If we were our brain, we would have full control over every thought that arose, yet we often experience intrusive, unwanted thoughts that seem to come from nowhere. Janet's story reveals how deeply ingrained familial narratives shaped her suffering. When she realized that much of her pain was not hers alone but rather a reflection of inherited fears and generational patterns, she began to liberate herself from their influence.

So, before we proceed further, let's confidently agree that...

You are not your brain!

Now, from a distance, let's explore your influence over suffering, starting with the choices you (and not your brain) make.

Embracing Agency

Fundamental to transforming suffering into joy is the knowledge that we have a choice in where we place our attention. The more we live in suffering (from a neurological perspective), the more we reinforce suffering's pathways in our brain. But the inverse is also true:

The more we direct our focus toward joy, gratitude, and appreciation, the less we suffer.

Suffering is rarely tied solely to the present moment. While a new trigger might seem to ignite pain, what it truly does is awaken old memories and ingrained narratives in the brain. These triggers tap into past experiences that are tightly woven into the fabric of our minds. For Janet, a conversation with someone about relationships or even an idle reflection could reawaken the pain of her failed marriage, flooding her with shame and self-blame.

What happens next is a shift to an internal *reaction*. Reactions are not deliberate; they are immediate, automatic, deeply rooted in the brain's subconscious patterns, and driven by survival. They run on well-worn neural pathways that are shaped by past experiences and dominated by the negativity bias. Once triggered, these reactions can drive further thoughts, fears, and behaviors that perpetuate the cycle of pain. The brain begins to replay and amplify the original pain, mingling it with the narratives we've constructed over time. The pain of the trigger is compounded by the old memories it reactivates, turning a fleeting moment into an ongoing state of suffering.

A *response*, on the other hand, is a deliberate, thoughtful action characterized by awareness, intentionality, and alignment with long-term well-being. In the context of pain, a response is a conscious choice to interrupt the pain-suffering cycle. It requires us to pause, recognize the moment of triggering, and actively decide how to move forward. This pause creates a moment of clarity in which we can challenge the old narratives and choose a different path. For Janet, a response meant acknowledging the pain of her triggered memory while reframing her narrative to acknowledge that everyone makes mistakes. This shift didn't erase her pain, but it interrupted her automatic reaction to it and began to weaken the grip of her fixed narrative.

Whereas a reaction enslaves us to the past, a response liberates us.

Choosing to respond intentionally rather than to react reflexively is an act of empowerment, enabling us to reclaim authority over the present and future. It is the conscious decision to step out of the shadow of old pain and into the light of possibility. With practice, we can transform the automatic reactions that perpetuate suffering into intentional responses that foster resilience and inner peace by enabling us to pause and ask, *What is the most constructive way to move forward from here?*

In order to begin to choose between reacting and responding, you need to remember that you are not your brain. Here are some simple and practical steps you can take to strengthen your agency over your brain and fully grasp the concept that you and your brain are two separate entities.

Pause mindfully. Before reacting to an emotional trigger, take a deep breath and pause. This disrupts automatic reactions and creates space for conscious choice.

Practice thought observation. Set aside time each day to sit quietly and observe your thoughts. Instead of engaging with them, simply notice them as passing clouds. This builds the skill of mental detachment.

Name your thoughts. When distressing thoughts arise, label them ("This is a fear-based thought" or "This is my brain predicting worst-case scenarios"). This helps reduce their intensity and reinforces the idea that they are separate from you.

Reframe narratives. When a painful belief surfaces, ask probing questions: Is this absolutely true? What evidence contradicts this? What would be a healthier way to view this situation?

Redirect your attention. Whenever suffering arises, deliberately shift your focus to something positive, whether it's gratitude, a small act of kindness, or a sensory pleasure like feeling the sun on your skin or savoring a sip of tea.

Build daily practices that support neuroplasticity. Engage in activities that reinforce new, positive thought patterns, such as journaling, meditation, visualization, or physical movement.

By applying these simple strategies, we take a crucial step toward controlling our minds rather than being controlled by them. The power of awareness, distance, and intentional redirection allows us to break free from the automatic grip of suffering and move toward a more deliberate and joyful existence.

Remember...

You are not your brain—
and you have the power to master it.

CHAPTER 5

Nurturing Hope

In this chapter, we take a deeper dive into the science of your brain—your *Brain Operating System (BOS)*—and how it shapes your experience of suffering. You'll learn practical tools to shift your thoughts, emotions, and attention in ways that reduce distress and build resilience. By mastering your brain's architecture, you nurture a more hopeful, peaceful way of being.

We turn now to explore the *Brain Operating System (BOS)* and the central role it plays in avoiding and averting suffering. By mastering the BOS, we can take conscious control over our internal world and build resilience in the face of adversity.

But to master it, we must first understand it. At its core, the BOS is the fundamental framework that governs how we perceive and react to the world around us. It dictates how we regulate and manage our thoughts, feelings, and fears, shaping the trajectory of our lives. Understanding the BOS enables us to shift from reactive, fear-driven actions to intentional, thoughtful responsiveness; gain control over negative thought patterns and emotions; cultivate resilience and adaptability in the face of life's challenges, without suffering; and enhance our capacity for joy, inner peace, and emotional strength.

Understanding the BOS

Let's begin by examining the BOS' five key components and their deliberate roles in ensuring our survival.

The ***prefrontal cortex (PFC)*** is the executive center of the brain, responsible for our awareness of our awareness, rational thinking and problem-solving, emotional regulation and impulse control, and decision-making based on long-term well-being rather than immediate reaction. Our PFC gives us the capability to observe our thoughts, feelings, and fears.

When we operate from our PFC, we gain clarity and perspective, enabling us to manage our magnificent brain. Through it, we make conscious choices, regulate our emotions effectively, and remain grounded even in difficult situations. Operating from the PFC is the key to resilience. It allows us to pause, reflect, and choose responses rather than being trapped in automatic reactions.

When stress, anxiety, or trauma hijack the brain, however, the PFC is often overridden by more primitive regions of the brain, leading to impulsive, emotionally charged decisions.

As the most primitive component of the brain, the ***reptilian brain*** resides at its base and is closely connected with the rest of our body through the spinal cord. It uses fear to drive our instinctual reactions, such as our fight, flight, or freeze mechanisms. These ancient neurological centers located in the brain stem are also responsible for basic survival functions like breathing, heart rate, and digestion.

The reptilian brain is critical for survival, but it operates without rational thought or emotional nuance. When we are triggered by perceived threat, this part of the brain takes over. Given the critical nature of its functionality, it has a very loud voice in our heads—as anyone who has been overcome by fear will testify. In its zeal to protect us, the reptilian brain drives many of the automatic, subconscious reactions that perpetuate suffering.

The ***cognitive brain*** is the brain's analytical powerhouse, responsible for learning, logic, reasoning, and language. Also known as the *neocortex* because of its recent development (in evolutionary terms), it processes information, solves problems, and enables rational decision-making.

When we avail ourselves of the clarity and perspective afforded by the PFC, we use data from the cognitive brain to make informed decisions and deliver considered responses. Without this executive override, our thoughts can become a jumbled noise, leaving us prone to relentless overthinking, spinning loops of worry, and self-criticism. Does this sound like suffering to you? I bet it does!

A Note to the Reader: You will see that I refer to thought as a universally positive phenomenon. Similarly, fear is always negative (even if it has positive outcomes, like saving your life in a dangerous situation). Another way of seeing this is to say that any negative thought is actually a fear. If you don't believe me, try thinking a negative thought that is not actually a fear in flimsy disguise.

The ***emotional brain***, sometimes referred to as the *limbic system*, is the part of the brain responsible for generating and processing emotions, particularly those related to fear, pleasure, and reward. It includes such key structures as the amygdala and the hippocampus.

The emotional brain is intimately attentive to the other centers in the brain, especially the reptilian and cognitive brains. It responds with great devotion to the loudest voice, which is why I call it the "great flip-flopper." So, when the voice of the cognitive brain prevails, emotions are positive. But, if the voice of fear rises and dominates, the emotional brain flips and suffuses our mind with negative emotions. This deep programing is both an asset and a curse.

While emotions are essential for human experience, the emotional brain can dominate our awareness when left unchecked. When combined with fear, negative emotions influence our thoughts and actions and can both precipitate and perpetuate unnecessary suffering. You probably recall how the amygdala's negativity bias makes us more likely to hold on to negative experiences than positive ones—an evolutionary survival mechanism that often leads to persistent fear, worry, and distress (in other words, suffering!). By understanding how the emotional brain interacts with other regions of the brain, we can learn to navigate emotions effectively rather than allowing them to control us.

The final component of the BOS is what I call the ***attention focusing consortium (AFC)***. It is the brain's director and gatekeeper of attention, managing what enters and stays in our conscious awareness. This consortium includes networks such as the *reticular activating system (RAS)*, which filters out irrelevant stimuli and focuses our attention on what we deem important. The AFC is crucial for directing our mental focus, making it a key player in neuroplasticity and suffering—because where attention goes, neurons grow. The AFC receives its instructions and hence is "programed" by the cognitive brain. This leaves it under our conscious control, if we choose.

The AFC is both the builder and the breaker of loops of suffering. When dominated by negativity bias, it continually spotlights painful memories and threats, reinforcing cycles of fear and sadness. Fortunately, when trained through practices like mindfulness, gratitude, and affirmations, the AFC redirects attention toward hope, joy, and possibility, accelerating the brain's ability to rewire and heal.

Mastering Your Brain: Systematic Resistance to Suffering

In essence, mastery of the BOS requires balancing the interrelated brain centers that compose it, ensuring that the rational cognitive brain leads us in *responses* to pain rather than the reptilian brain that instinctively triggers thoughtless *reactions*. This mastery is achieved by embracing the privilege of perspective granted by the PFC, and by reprograming our AFC to direct our attention toward information that serves rather than undermines our intentions.

A common misconception is that cultivating a positive mindset requires that we ignore pain. In reality, it involves choosing to move beyond it. Resilience is not about avoiding adversity—it is about developing the capacity to recover from, adapt to, and emerge stronger in the face of difficulty or protracted pain. We may consider resilience to be the property we exhibit when we can successfully escape pain and difficulty without becoming entrapped by suffering.

Mastering the BOS enhances resilience by equipping the brain to navigate uncertainty and setbacks with greater ease. It shifts attention away from suffering and toward constructive solutions, fostering a mindset rooted in growth and possibility. Through consistent practice, BOS mastery strengthens emotional intelligence and self-regulation, enabling us to manage our responses to challenges more effectively. Additionally, it supports the rewriting of limiting mental narratives, replacing them with an empowered, solution-focused outlook that promotes long-term well-being.

Mastering your BOS is not an abstract concept—it is a practical, trainable skill that offers freedom from unnecessary suffering. Understanding the distinct roles of the critical elements of your brain and learning to operate them effectively grants you the ability to shift from reactivity to responsiveness, from suffering to resilience, and from chaos to inner peace. In the next sections, we will explore the practical applications of BOS mastery to systematically train your brain to operate in its most powerful, balanced, fulfilling, and suffering-resistant state.

Pause First: Opening the Mind

One of the most useful tools available to us in overcoming suffering is the ability to create a gap between stimulus and response. Without this gap, we react impulsively, dictated by the emotional and survival-driven parts of the brain. However, by harnessing the power of the PFC, we can train ourselves to pause, reflect, and choose our responses intentionally rather than being enslaved by automatic reactions.

A powerful method to engage your PFC and ensure you are operating from a place of calm, considered reflection is to ask three critical questions:

> ***What am I thinking?***
>
> ***What am I feeling?***
>
> ***What am I fearing?***

If you can answer these three questions accurately, you are seeing the world from your PFC. By identifying your current thoughts, you shift from unconscious reactivity to conscious observation, allowing you to examine your underlying narratives and assumptions with clarity. By acknowledging your emotions without judgment, you differentiate between transient feelings and deeper truths. By recognizing your fears, you confront them directly and assess whether they are rational or exaggerated.

Mastering these three questions is both diagnostic and developmental: They serve as proof that we are operating from the PFC and simultaneously train us to build greater self-awareness and emotional regulation. Through consistent practice, we gain increasing competence in navigating challenges from a place of clarity, control, and resilience—dramatically reducing unnecessary suffering.

The ability to operate from your PFC is a cornerstone of BOS mastery.

Although the three questions are the root of PFC operation, I have a process to help more fully guide you to consciously engage your PFC. These five deliberate steps leverage the unique functionality of the BOS to transition from reactive patterns to empowered responses.

Step 1: Begin with a brief, mindful pause. An intentional pause shifts your awareness into your PFC, creating space between stimulus and response. This brief moment sets the stage for self-awareness and primes your brain for the internal journey ahead. Center yourself with three deep breaths, grounding your attention in the present moment before proceeding.

Step 2: Survey your emotions to feel negative emotions. Tune in to your emotional state. Focus on a specific challenge or problem, and ask yourself, *How do I genuinely feel about this?* Observe and name the emotions that surface, paying attention to any physical sensations in your body. Pay particular attention to any negative emotions you feel. This step engages the emotional brain and begins to separate your identity from your feelings.

Step 3: Identify and acknowledge your fears. Shift your attention to the source of your discomfort. Identify all the fears driving your emotional reactions. Use probing questions to uncover both conscious and unconscious fears: *What am I afraid of? What are the possible outcomes and consequences? Who else is affected, and how?* Probe persistently, pressing through surface-level fears until you reach the root causes. Bring the strong flashlight of courage in the deep reassurance that bringing fear into the light reduces its power and activates the PFC's problem-solving capabilities.

Step 4: Notice positive thoughts emerging from clarity. Shift your attention to your cognitive brain, seeking clarity through questions such as, *What do I think about this situation? What is the real story? What perspectives would others offer?* This is where cognitive override takes place; the PFC assumes control, guiding you to choose rational responses over emotional reactions. As your fear settles and clarity emerges, observe the natural shift toward positive thoughts and solutions. If negative emotions resurface during this step, return to Step 3 to uncover any newly identified fears before continuing.

Step 5: Survey your emotions again to feel positive emotions. Complete the process by reconnecting with your emotional state. Reflect on the positive emotions emerging from your new, empowered thoughts. Ask yourself, *What positive emotions am I experiencing? Where do I feel them in my body? What empowering beliefs are being fueled by these positive emotions?* Notice and celebrate the positive emotions—such as relief, hope, or gratitude—that naturally follow deep thoughtfulness and conscious reflection. If negative emotions arise, acknowledge them as signs of unresolved fears and return to the previous steps. Repeat the process until positive emotions fully replace the negative ones, marking the triumph of your cognitive and prefrontal leadership over fear and emotional turbulence.

This five-step process is the key to fully activating the PFC and assuming control of your BOS. Practiced regularly, this process puts you solidly in control, shifting you from reactive patterns to conscious, empowered responses, and will be a cornerstone of your work to liberate yourself from unnecessary suffering.

And, while the five-step process is the most powerful tool I know to train yourself to operate from your PFC, there are several other simple ways that we can powerfully cultivate presence.

Pause and reflect. In the heat of the moment, even a brief pause can interrupt suffering and reengage the PFC.

Pause and breathe. Take three slow, deep breaths, inhaling through your nose and exhaling through your mouth. This helps regulate the nervous system and shift your focus to the present.

Anchor to a physical sensation. Feel the weight of your feet on the ground, notice the temperature of the air, or focus on the rhythm of your breath.

Count backward from 10. This simple mental reset allows your PFC to regain control in heated or distressing moments.

Name the emotion. Saying "I am *experiencing* frustration" rather than "I *am* frustrated" creates distance between the emotion and your identity, reminding you that you are not your brain.

Listen deeply. Give the gift of deeply engaged listening, focusing entirely on the other person without preparing a response. Perhaps you remember how, as an older woman, Janet chose to be present with and listen fully to her son...a life-changing moment that made a world of difference for her.

Let go of self-interest. Fully commit to the here and now without agenda or judgment.

By learning to stay present, we break free from the cycle of suffering and move toward calm, intentional action. Through deliberate practice, we can learn to pause, reflect, and redirect our minds toward clarity, resilience, and joy. By consistently applying these principles, we shift from automatic suffering to conscious living—from being trapped in reactivity to fully embracing the power of choice.

Mindful Presence: Awareness Rules

A deep state of presence is sometimes referred to as *mindfulness*. I like to write it out as *mind-full-ness*, reminding me that the mind is always full and that, even (perhaps especially) when I am captivated by fear and negative emotion, I should be searching my mind for other things, like thoughts and positive emotions. I can only do this from my PFC.

Mindfulness has been well researched and proven to be a powerful tool for shifting from emotional reactivity to rational responsiveness. Practicing mindfulness has been shown to rewire the brain, strengthening the PFC and weakening the grip of the amygdala (the center in the emotional brain that is so highly susceptible to fear). Mindfulness also shifts our focus from dwelling on the past or fearing the future to remaining present, which helps prevent unnecessary rumination and anxiety. By integrating simple mindfulness techniques into daily life, we systematically train our BOS to function from a place of control rather than chaos, fostering both resilience and inner balance.

Remarkably, dedicating just five minutes daily to mindfulness meditation can initiate profound neurological changes, fostering greater emotional regulation and inner peace.

Try this practical mindfulness exercise. Sit comfortably, and close your eyes. Inhale deeply through your nose for four counts. Hold the breath for four counts. Exhale slowly through your mouth for six counts. Repeat the pattern for two minutes, focusing solely on your breath and calmly redirecting your mind back to your breathing if it wanders.

The more we engage in mindfulness, presence, and intentional action, the less suffering controls our experience. When we operate from our PFC, we are grounded, deliberate, and capable of responding to life's difficulties with wisdom rather than reactivity.

Cognitive Override: Thoughts Shape Reality

The irony of suffering is that it both originates from and finds resolution within the cognitive brain. Unique to humans, this part of the brain grants us memory and imagination—remarkable faculties that, when hijacked by fear, fuel suffering. It is rarely the present moment that causes our distress; rather, it is the mind's replay of past wounds and its anxious projection of future pain that entrap us in cycles of suffering. Fear grips our memories and contaminates our imagination, distorting our reality and chaining us to unnecessary anguish.

Yet, within this same cognitive machinery lies liberation from suffering. The cognitive brain exerts hierarchical dominance over fear, enabling us to suppress fear with rational thought. It allows us to view memories clearly, free from emotional distortion, and empowers us to override fear-driven narratives with solution-oriented reasoning. Mastering the liberating process of cognitive override is not only possible; it is essential to breaking free from suffering's grasp.

Cognitive override operates on the principle that thoughts dictate emotions, not the other way around. More than this, thoughts (which are always positive) elicit only positive emotions. Sounds good, right? It certainly feels good! When we choose how to think about a situation, we influence not only our feelings but also our behaviors and physiological state.

By deliberately reshaping our thoughts, we can alter the emotional and physiological reactions that follow, breaking free from cycles of unnecessary suffering. This is not about suppressing emotions but rather about consciously influencing the nature of our prevailing emotions by redirecting our mind toward thought. Through reinterpreting past experiences in ways that reduce emotional turmoil, we reclaim control over our inner narrative. This practice builds mental resilience as we actively choose thoughts that promote confidence and well-being. Over time, it fosters the creation of new neural pathways, reinforcing positive cognitive patterns and setting the foundation for long-lasting emotional growth.

One of the most well-documented applications of cognitive override in clinical psychology is *cognitive behavioral therapy (CBT)*. CBT is a structured, evidence-based approach that helps individuals identify and challenge negative thought patterns (what I call "fears"). While originally developed for clinical use, its principles can benefit everyone. The core premise of CBT is that our thoughts influence our emotions and behaviors—so, by changing our thoughts, we can change our experience of life. And there are some key CBT techniques that facilitate cognitive override.

Engage in cognitive restructuring. Examine your fears, and replace them with balanced, constructive perspectives (thoughts).

Try thought journaling. Write down your fears, and challenge their validity to break automatic negative-thinking loops.

Activate positive behaviors. Engaging in positive actions, even when your emotions resist, shifts your mental state toward optimism and productivity.

Apply reality testing. Ask yourself, *Is this thought absolutely true?* to separate facts from exaggerated fears.

By adopting these techniques, we interrupt suffering while regaining control over our minds and emotional landscapes.

Many people unknowingly reinforce their own suffering by holding onto distorted interpretations of past events. When they string these fears and negative assumptions together, they build a destructive story that fuels suffering. One of the most transformative applications of cognitive override is known as *cognitive reframing*, or shifting the narrative we hold about our prior experiences. When we reframe a negative narrative to a positive one, by applying cognitive override, we change its impact on our emotions and future responses.

Initially, Janet viewed her past through a lens of self-blame and inadequacy, reinforcing a cycle of suffering. Through reflection and guidance, however, prompted by her discussion with Hope, she began to reframe her past pain as a source of growth—allowing her to cultivate self-compassion and resilience instead of continued suffering.

To apply cognitive reframing, step back from automatic interpretations. Ask, *What is another way to view this experience?* Separate facts from emotional embellishments. Challenge sweeping statements like "I always fail" or "I am unworthy." Then, rewrite your personal narrative. For example, instead of saying, "I was abandoned," Janet ultimately reframed her narrative to be, "I survived a painful experience and learned strength and compassion."

The ability to consciously override negative thought patterns requires consistent training. Like strengthening a muscle, we must exercise the mind's ability to choose empowering perspectives over automatic suffering. There are several daily habits to cultivate cognitive override.

Set a positive cognitive tone for the day through morning affirmations. Affirmations are a powerful way to set intentional thought patterns from the moment we wake up. By consciously choosing affirming beliefs, we override subconscious fears and self-doubt. Start each day by repeating three empowering affirmations that challenge negative thought patterns (e.g., "I am strong, capable, and worthy"). Use a mirror technique—speak affirmations aloud while looking at yourself, reinforcing belief in your words. Pair affirmations with deep, intentional breathing to strengthen their neurological impact.

Reinforce resilience by setting incremental goals. The process of overcoming suffering and retraining thought patterns does not happen overnight. Small, deliberate actions compound over time, leading to significant transformation. Set one small goal per day that reinforces positive change (e.g., practicing mindful breathing for five minutes, rewording one negative thought into a constructive one, or engaging in an act of kindness). Then, acknowledge your progress—each step forward, no matter how small, is a reinforcement of cognitive mastery. And when you reach your goals, celebrate those achievements; reinforcing success through positive acknowledgment strengthens the PFC's role in guiding thoughts and actions.

The greatest power we hold over our suffering
is the ability to choose our thoughts.

By practicing cognitive reframing, engaging in gratitude and affirmations, and setting intentional goals, we take command of our internal dialogue, transforming suffering into personal mastery. The journey begins with a single realization:

We are not our thoughts,
but we have the power to shape them.

Emotional Regulation: Harnessing Feelings

The ability to regulate emotions is central to reducing suffering and cultivating inner peace and joy. Our emotional brain is wired to react instantly, prioritizing threats and amplifying distress signals. In contrast, emotional regulation is the skill that allows us to take a step back, process emotions with awareness, and choose responses that align with our long-term well-being. By understanding how emotions arise and employing strategies to manage them, we reclaim control over our internal experiences.

Janet's early experiences of abandonment and pain created deep neural pathways that made her highly sensitive to perceived rejection. Even in safe situations, her brain would interpret neutral or ambiguous signals as threats, triggering intense emotional reactions. The dense haze of negative, fear-based emotion clouded Janet's perspective; past pain overrode present reality, keeping her trapped in cycles of suffering.

An unchecked negative emotional state traps us in a loop of suffering, fueling endless rumination where past pain replays as if it were unfolding in the present. It heightens fear reactions, prompting us to anticipate future suffering through the distorted lens of past experiences. This turmoil blurs the line between real threats and conditioned emotional echoes, leaving us reactive and vulnerable. The encouraging truth, however, is that the brain can be trained to override these patterns, calming the emotional noise, breaking the cycle, and restoring clarity and balance.

Name them to tame them. Research has shown that naming the emotion (e.g., "I am *feeling* anxious" rather than "I *am* anxious") induces a shift to thought-full-ness and cognitive override and reduces the amygdala's amplifying influence. This directly and indirectly helps us to process emotions from a place of clarity and reason rather than fear-driven impulsivity.

So, when a strong emotion surfaces, pause and take a deep breath (invoking your PFC). Identify the emotion, and name it directly: *I feel overwhelmed*, *I feel frustrated*, or *I feel vulnerable*. Without judgment, observe its presence in your body—tightness in the chest, a racing heart, or tension in the shoulders. Finally, remind yourself that emotions are transient: *This feeling will pass.*

Through consistent practice, emotion labeling cultivates self-awareness, reduces reactivity, and breaks the cycle of suffering-inducing thought patterns. And once that cycle is broken, we can introduce new emotional patterns that promote resilience and joy. The most powerful of these practices is gratitude. Far more than a fleeting feeling, gratitude is a powerful neurological tool that reshapes our mental landscape and offers a strong antidote to suffering.

Research shows that consistent gratitude practice rewires the brain, fostering new neural pathways that counteract our natural negativity bias. Functional MRI studies reveal that practicing gratitude enhances activity in the PFC. Repeatedly focusing on appreciation strengthens neural circuits that default to positivity rather than distress.

Even better, regular expressions of gratitude boost dopamine and serotonin—the brain's natural feel-good chemicals—promoting feelings of joy and well-being. By shifting focus from what is lacking or painful to what is positive and present, gratitude becomes a pathway that leads from suffering to inner peace. Additionally, gratitude cultivates emotional resilience, enabling quicker recovery from distress and reducing symptoms of anxiety and depression. With regular practice, as in the suggestions below, gratitude transforms the brain's responses, creating a lasting foundation for inner peace and joy.

Keep a gratitude journal. Start by recording three things you are grateful for each day. Focus on specific moments, such as "the warmth of the sun on my face" or "a friend's comforting words." Reflect on why each experience matters to you. Over time, this practice reprograms your brain to seek positivity, gradually shifting your emotional baseline from suffering toward joy.

End each day on a positive note by reflecting on gratitude before sleep. Before going to sleep, recall one positive experience from your day. Relive the moment in your mind, allowing yourself to feel the joy or comfort it brought. This simple reflection reduces stress and reinforces positive emotions, helping you end your day with gratitude.

Strengthen bonds and increase joy by expressing gratitude to others. Take time to verbalize your appreciation to others. A sincere "thank you" or a short note of gratitude can deepen relationships and elevate your mood. Expressing appreciation activates neural pathways linked to joy, benefiting both you and the recipient.

Embrace the present moment with gratitude walks. Incorporate gratitude into your daily routine by taking mindful walks. As you walk, consciously observe the world around you—the rustling leaves, a gentle breeze, or a fleeting sunset. This practice combines mindfulness with appreciation, enhancing your awareness of life's simple pleasures.

Find joy in everyday moments. Pause during your day to savor small moments of joy, such as the taste of your coffee, a pleasant conversation, or a quiet moment of stillness. Brief acknowledgment of these everyday gifts helps build a resilient, gratitude-centered mindset that counteracts negativity and suffering.

Gratitude offers a profound path from suffering to inner peace and joy.

Gratitude leverages the brain's capacity to rewire thought patterns toward positivity. Initially lost in self-blame and despair, Janet discovered that acknowledging simple joys—like acts of kindness, shared connection, and moments of beauty in her garden—began to shift her perspective. Guided by Hope's spontaneous and heartfelt expressions of appreciation, Janet gradually retrained her brain to seek light in the midst of darkness.

And that leads us to the ultimate endpoint of emotional regulation: *self-compassion*. At the root of much suffering lies self-criticism. Harsh self-judgment triggers negative emotions and perpetuates cycles of pain; this, by definition, becomes suffering. Self-compassion—the practice of offering oneself the same kindness extended to a friend—disrupts this destructive pattern. For years, Janet carried the burden of guilt and self-blame, which compounded her suffering. However, as she began to embrace self-compassion, acknowledging her pain without judgment, she discovered a path to emotional freedom and healing. You can do the same through the suggestions below.

Acknowledge your suffering without judgment. Shift your inner dialogue from *I am weak for feeling this way* to *I am experiencing a moment of struggle*.

Practice self-kindness. Replace self-critical thoughts with affirming ones, such as *I am doing my best* or *I deserve care and understanding*.

Recognize shared humanity. Remember that suffering is a universal experience; you are not alone in your challenges.

By adopting these practices, we soften the grip of self-criticism, fostering inner peace and resilience. Practical application of these techniques strengthens our ability to pause, process, and choose our responses, ensuring that emotions serve us rather than control us. With deliberate effort, we move away from emotional turbulence and toward emotional mastery—one moment at a time.

Manifestation: Programming the AFC

Manifestation is often mischaracterized as a mystical concept detached from reality. Its excessive use in poorly substantiated "new-age" guidance has given it a very bad reputation. When understood through the lens of modern neuroscience and the BOS, however, manifestation becomes a credible, scientifically supported process. It is the alignment of focused intention, directed attention, solid belief, and actionable behavior to drive real, measurable change. This alignment is achieved through the deliberate use of our cognitive, emotional, and attentional resources—as outlined above—to shape outcomes and influence reality.

Visualization, a key component of manifestation, engages neural circuits that support goal achievement, confidence, and resilience. When we repeatedly create a mental image of a desired future, we activate the brain's reward pathways, reinforcing motivation and programming the BOS to seek opportunities aligned with our intentions. By integrating cognitive and emotional perspectives—imagining not only the outcome but also how it will feel to achieve it—we create powerful neural imprints that prime the brain for success.

Intentional focus further enhances manifestation by engaging the AFC. By directing the AFC toward our desired future, we begin to notice opportunities, resources, and solutions that were previously invisible. This process effectively channels the brain to act in accordance with our goals, driving decision-making and behaviors that lead to their realization.

The Journey Continues

Master your BOS, and you reclaim your power to suffer less and live more.

Having explored the inner workings of your BOS and the practical tools it offers, you now hold the keys to interrupting the patterns that perpetuate suffering. While these strategies form the foundation of transformation, our journey together continues on from here, if you choose to pursue it. In the next chapter, we'll deepen the work to help you move beyond surviving toward truly thriving.

CHAPTER 6

Sustaining Hope

In this chapter, we examine how to sustain your transformation by addressing its deeper roots. You'll learn how inherited narratives and unresolved pain from the past shape our inner world and how to break that cycle for yourself and future generations. I'll introduce the concept of the patchwork mind, showing how your sub-personalities influence your behavior and how reclaiming your authentic SELF can restore inner harmony. Finally, we'll explore forgiveness as a powerful means to release the lingering weight of suffering and create space for enduring peace.

Now that you possess a deep understanding of your BOS and know how to master it to overcome suffering, I'd like to address a few of the more advanced strategies to combat deeply embedded suffering. These strategies will help you develop inner peace and joy, sustaining hope throughout your life.

Breaking the Cycle of Inherited Suffering

The transmission of suffering across generations is a deeply intriguing phenomenon, one that extends beyond shared stories and familial habits. In the preceding sections, we have illustrated the inherent predisposition of the human brain toward suffering. But beyond this universal inclination lies a more nuanced phenomenon: the generational transmission of a "suffering disposition." This inherited mindset often manifests across generations, embedding itself deeply in family systems and communities. The question is: How does this transmission occur?

The most apparent mechanism is through learned behavior. Just as children absorb language and social customs from their caregivers, they also internalize patterns of emotional reaction. If a child grows up watching a parent respond to life's challenges with anxiety, despair, or resignation, the brain encodes these patterns as standard operating procedures, unconsciously believing that, if a respected adult does it, it must be a successful habit worth adopting.

The parent's suffering becomes the child's emotional blueprint, laying the groundwork for a lifetime of habitual suffering.

Beyond learned behaviors, shared environmental experiences contribute to the generational perpetuation of suffering. Family units often navigate common external stressors—economic hardship, societal discrimination, or traumatic events—that collectively shape their emotional narratives. These experiences create shared neural imprints that reinforce a collective suffering mindset. For instance, the well-documented cycle of abuse exemplifies how trauma can reverberate across generations. Children who experience or witness abuse often carry these emotional scars into adulthood, perpetuating the cycle by unknowingly replicating the behaviors they once feared. The abused child becomes the abusing parent, and the narrative of suffering continues its relentless march through time.

The field of epigenetics has provided further compelling insights into how trauma imprints itself across generations. Emerging research suggests that both genetic inheritance and epigenetic modifications can perpetuate a predisposition to suffering. While specific genes do not encode a "suffering disposition," certain genetic variations can heighten susceptibility to anxiety, depression, and emotional distress.[130,131] These vulnerabilities, when paired with environmental stressors, shape the mind's narrative of suffering and influence emotional regulation.

For instance, a study on survivors of the Tutsi genocide in Rwanda demonstrated how extreme psychological trauma left measurable epigenetic marks on both survivors and their children.[132] These changes were associated with the stress-response systems of the body, suggesting a biological inheritance of trauma beyond lived experience.

The implications are myriad. Trauma and suffering appear to ripple through generations not merely through behavioral modeling but through biological adaptation. This helps explain why some patterns of suffering feel deeply embedded. And while the mechanisms are still being unraveled, the science underscores a hopeful truth: If suffering can be biologically transmitted, then intentional practices like mindfulness, gratitude, and emotional regulation might similarly reshape the narrative for future generations.

Numerous individuals have broken free from these cycles through therapeutic intervention, mindfulness practices, and the conscious cultivation of healthier emotional habits. Understanding the powerful influence of learned behavior and shared experience equips us to recognize the origins of our suffering and to consciously disrupt the patterns that no longer serve us.

Janet had fully raised her children by the time she met Hope. It is my determination in my work, and through this book, that we empower young parents to liberate themselves from their suffering before they raise their children—even if it turns out that there are strong biological inheritance patterns.

Breaking the cycle of inherited suffering requires intentional, sustained effort to rewire deeply ingrained neural patterns and foster healthier emotional habits. Several specific focuses for inherited suffering are worth considering.

The first step in breaking the generational cycle of suffering is to recognize and challenge the narratives we've inherited. These stories, often absorbed without question in childhood, shape how we perceive ourselves, others, and the world around us. Reeducating the mind begins with cultivating awareness: recognizing the internal dialogue and identifying beliefs rooted in fear or negativity. For example, if someone grew up hearing phrases like "life is always a struggle," they can pause, challenge this belief, and reframe it with an empowering alternative such as, "Life presents challenges, but I have the resilience to navigate them." Through deliberately introducing alternative perspectives and by repeatedly questioning and reframing limiting beliefs (aka fears), we activate the cognitive brain and gradually move away from our suffering.

The next step is to build a new legacy for future generations, one grounded in hope, resilience, and gratitude—starting with you. This process involves consciously modeling emotional resilience, practicing self-compassion, and reinforcing the principles of emotional well-being through daily interactions. Gratitude practices, for instance, can serve as powerful anchors for positive emotional states. Encouraging family members to share daily moments of appreciation, whether at the dinner table or during evening reflections, gradually shifts the collective focus from suffering to joy. As Janet embraced new narratives of self-compassion and learned to appreciate the present, she not only alleviated her suffering but also set a powerful example for those around her.

Breaking the cycle of generational suffering requires courage, consistency, and conscious effort. Yet, with these strategies, we can rewrite our mental narratives, transforming inherited pain into a legacy of hope and emotional strength.

Deep Healing to Liberate the SELF

In my individual work with clients, I have adopted and adapted insights from many of the greatest thought leaders in the fields of psychology, psychiatry, and the associated social sciences. Using the work of Sigmund Freud, Carl Jung, and the Gestalt theorists, as well as more modern approaches like Voice Dialogue pioneered by Hal and Sidra Stone and the Internal Family Systems model of Richard Schwartz, I employ a phenomenon I refer to as *psychic polymorphism*.

Neuroscientists today agree that the most remarkable aspect of the human brain is its extraordinary adaptability and versatility. We achieve this versatility as a consequence of our inner heterogeneity—a vast, perhaps limitless array of inner "algorithms" that make us each who we are. Different from all the other 8 billion living human beings, we each have our own unique "psychic fingerprint."

How is this possible, that each one of us can be totally different from all other human beings? It relates to the seemingly limitless adaptive capacity of the human brain.

When we are born, we are estimated to have between 80 and 100 billion individual nerve cells (called *neurons*) in our brain. We know that each neuron connects to many other neurons, playing its own singular role in the computational circuit that is formed. Within this unimaginable lattice of connectivity, it is estimated that we have over 100 trillion connections, each contributing to our own unique neurobiological complexion.

For the first two years of our life, the brain is connecting these neurons at a voracious pace. We believe that over a million connections are made per second during this period.[133] This leaves us with incomprehensibly vast computational capability by the age of two. And then, a very interesting thing happens: To limit the unsupportable energy demands of such a huge biological supercomputer, to facilitate reasonable mental focus, and to allow us to adapt based on our particular context, we start breaking down the less-useful connections.

It is this process that results in our individual personality, which by now you may realize is not in fact a singular, homogenous entity but a vastly complex composite of a myriad of sub-personalities—each algorithm effectively supporting a separate thought-attitude-behavior pattern. It is this collection of sub-personalities that forms who we are and gives rise to my term *psychic polymorphism*. "Poly" means many, and "morph" means shape. So, our psyche is formed by the aggregation of many shapes (or sub-personalities).

Why is this important to suffering? Because you can see that the algorithms of our brain (or the connections of neurons referred to as *sub-personalities*) develop in response to the unique stress, distress, and trauma that we face as a developing individual. Not surprisingly then, these specific algorithms not only encode memories of the most painful events of our lives but also perpetuate the dysfunctional attitudes and behaviors that result from them, making them significant agents in generating and nourishing suffering.

When you think about some of your own sub-personalities, such as your tireless people-pleaser, your striving perfectionist, or your busybody taskmaster, you will begin to recognize how some of them are driven by fear and trigger intense negative emotions. They may have been very useful in childhood—nobody will argue that working hard is a good thing. But when you realize that you are addicted to work simply to prove your value to the world, you will realize that your "taskmaster" is now foundationally active in precipitating and perpetuating your suffering. You will also notice that each of these fear-driven sub-personalities is supported by a false narrative, for example, the belief that only perfection is good enough for you to live a successful and happy life. On the contrary, perfectionism—when it dominates your behavior, influencing your thoughts, emotions, and actions beyond their initial protective intent—is a dysfunctional strategy that results in unnecessary human suffering!

The good news is that, beneath the swirl of competing sub-personalities that populate your inner world—each shaped by fear, survival, and past experience—there exists a singular, unwavering presence: YOU. This authentic SELF is the conscious observer, the wise integrator, and the seat of clarity and calm. While your fear-based parts react reflexively to the world, the authentic SELF responds with discernment and purpose.

It is from this center that you can witness your suffering without becoming it.

Choose your path rather than being driven by habit, and gently lead your internal system toward healing and wholeness. As you deepen your awareness of this SELF, you awaken the most powerful ally in your journey from suffering to peace.

When your constellation of sub-personalities (I often refer to them as your "board of directors") takes over your life—regarding you as their servant, rather than the other way around—you have effectively subjugated yourself to your magnificent, survival-based brain. And so, you suffer unnecessarily. On the other hand, when you realize that your brain is the most incredible gift you will ever receive and is designed to serve your authentic SELF, and you take the steps required to master your relationship with your brain, then you have exercised your right to escape unnecessary suffering.

It is difficult to do this work alone, and I always encourage people to seek out an appropriately qualified professional coach, counselor, or therapist to help. If you are experiencing meaningful disruption from these algorithms, you should always seek the guidance of a certified clinical practitioner. For the rest of us, it may be enough to accept an invitation to journey with me into the theatre of your mind, deploying your imagination—which, ironically, you have learned is a problem in precipitating suffering—to assume authority over the algorithms. In this powerful theatre, you can employ visualization techniques to project and interact with your sub-personalities, facilitating detachment, understanding, and eventually mastery over these automatic reactions.

The simplest and most powerful step toward adopting this approach is to start using the word part. So, you will say, "A *part* of me believes that life is harsh and frequently causes me pain," rather than pronouncing the vastly pervasive and permanent edict that "Life is harsh." If I am standing close by, I will ask you, "And what does the rest of you think?" You will realize that somewhere deep inside you is a little voice that says, "Life is beautiful and richly rewarding." When you hear that voice, I will ask you which narrative you choose to live by. If you choose the latter, you will have taken a meaningful step away from unnecessary suffering.

Using the principles of psychic polymorphism, you can ultimately empower yourself to transcend survival-driven behaviors, fostering a state of thriving and self-mastery by reclaiming authority over your mind's inner landscape. By working with, rather than against, these fear-driven parts, you can achieve greater self-awareness and emotional regulation.

This exploration of psychic polymorphism underscores the intricate and fragmented nature of the human mind—a complex interplay of protective algorithms designed to ensure survival but often contributing to unnecessary suffering. Understanding these patterns provides the foundation for deep, lasting healing.

By recognizing, engaging with, and ultimately reprogramming these fear-based sub-personalities, we empower ourselves to reclaim mastery over our emotional landscape.

The journey away from suffering is not about erasing these protective parts but about learning to relate to them with compassion, curiosity, and conscious authority.

Forgiveness: Severing the Ropes That Bind Us

One of the ultimate steps in recovering from suffering is forgiveness, both of ourselves and of others. This forgiveness plays a crucial role in severing the emotional ties that bind us to past trauma. Studies in neuroscience suggest that, when we hold on to resentment, our brain remains in a heightened state of stress, activating the amygdala and reinforcing pathways of pain and distress. Letting go of these emotional burdens, on the other hand, engages the PFC. As we grasp agency, through forgiveness, we release the physiological and psychological weight of past wounds.

Practicing forgiveness does not mean excusing harmful actions or denying the pain experienced; rather, it is a conscious decision to free ourselves from the continual reexperiencing of that pain. Practical strategies for fostering forgiveness include writing letters (without the need to send them), engaging in guided self-compassion meditations, and using cognitive reframing to shift perspective. Research suggests that individuals who actively practice forgiveness experience lower cortisol levels, improved cardiovascular health, and greater overall well-being. By embracing forgiveness, we reclaim agency over our past, transforming trauma from a lingering wound into a source of growth and strength.

As she told her story to Hope, Janet was clearly forgiving her father for going to war, the train for taking her away from her home, the nurses for taking her mother away, and her mother for abandoning her. Although not stated, we could even sense Janet forgiving herself for the discord her suffering brought into her own marriage and family. Her ability to finally forgive and let go freed Janet to find peace instead of continuing in her suffering. I encourage you to do the same.

Owning the Transformation

As we conclude these chapters on transforming suffering into strength, we arrive at a profound realization: The journey of healing does not end with the cessation of suffering; that is a lifelong process that evolves with each new experience, offering continuous growth and deeper self-awareness. Rather, the journey marks the beginning of a more purposeful, hope-filled existence. We have explored the intricate workings of the brain, the influence of inherited suffering, and the transformative power of conscious engagement with our mental processes. The tools and insights presented here demonstrate that suffering is not a fixed condition but a malleable experience shaped by the narratives we choose, the practices we adopt, and the neural pathways we reinforce.

This journey requires persistent effort and compassionate self-awareness. The strategies we have examined—from cognitive override to emotional regulation, from gratitude practices to self-compassion—serve as guideposts for navigating life's inevitable challenges without succumbing to unnecessary suffering.

Yet, the ultimate goal extends beyond the self. As we learn to transcend our suffering, we naturally radiate hope to those around us. In this way, personal healing becomes collective healing, rippling through our relationships, families, and communities. Janet and Hope remind us of the generative power of compassion and the contagious nature of courage. When one person awakens to their capacity for peace, they illuminate a path for others to follow. The same principles that allow us to transform suffering also empower us to build a future brimming with promise, connection, and vitality. Now, let us step into the light of that hopeful horizon.

CHAPTER 7

A Legacy of Hope

The journey from suffering to joy is neither linear nor finite. It is a dynamic, evolving process that unfolds across the canvas of our lives, often in ways we least expect. Transformation is not a destination but a lifelong commitment to growth, resilience, and inner peace. As we apply the disciplines we explored in the previous chapters—mastering the BOS, engaging cognitive override, practicing emotional regulation, and cultivating gratitude, among others—we set ourselves on this hopeful path.

The path to hope requires that we embrace both its continuity and its imperfection. Emotional healing defies the neat, linear progression we might wish for. Much like the ocean's tides, it ebbs and flows, sometimes returning us to familiar pain before propelling us toward newfound peace. This nonlinear nature can be disheartening if misunderstood, yet it is an essential part of the process. Each challenge presents an opportunity to deepen our practice and reinforce our capacity to choose peace over suffering.

Moments of suffering will continue to arise—this is the nature of the human experience. But when we engage our BOS with intentional practices like mindfulness, gratitude, and self-compassion, we learn to navigate these moments with grace. These tools help us soften the impact of distress, shift our focus toward possibility, and restore our innate sense of hope.

As Janet revisited her memories of abandonment and loss, she experienced waves of sadness. Yet each time she faced these emotions with curiosity rather than resistance, she uncovered new insights. With each cycle, the pain lost some of its power, making space for hope to take root and grow.

Hope, like joy, multiplies when it is shared. Through her interactions with Hope, Janet not only found personal peace but also became a beacon of possibility for those still navigating their own pain. When we extend our journey beyond ourselves—when we share our insights, struggles, and triumphs with others—we contribute to a broader legacy of healing and resilience.

Human connection is essential to this process. Suffering often isolates us, convincing us that we are alone in our pain. When we share our stories within a compassionate community, we dissolve that delusion. We discover that suffering is universal, and so is the potential for healing.

As you walk your own path toward hope, consider the ripple effects of your journey. Just as Janet's willingness to engage with her suffering helped Hope find clarity and courage, your resilience can also inspire others. When you choose to approach challenges with mindfulness, when you reframe negative narratives with intention, and when you cultivate gratitude amidst life's inevitable struggles, you become a living example of hope in action. Your transformation does not end with you; it becomes a gift to those you encounter: family members, friends, colleagues, and even strangers.

Individual healing contributes to collective well-being. As you develop greater emotional agility and resilience, you influence your environment in subtle yet profound ways. Acts of kindness, expressions of gratitude, and demonstrations of compassionate presence create a positive feedback loop, encouraging others to explore their own capacity for healing.

Teaching, paradoxically, is one of the most effective ways to solidify our learning. When we articulate our experiences to others, we deepen our understanding and reinforce our new mental patterns. The student who becomes the teacher continues to grow as a self-healer. Janet's willingness to share her experiences with Hope did more than offer the young woman guidance—it became a catalyst for her own healing.

At times in our story you may have found yourself asking a question: Is Hope healing Janet, or is Janet healing Hope? In fact, both the speaker and the listener are often healed in the same conversation. You will be astounded by how much you will learn when you spread your new knowledge about suffering and hope.

Take a moment to craft a personal legacy statement—a declaration of the values, principles, and emotional states you wish to embody and pass on to others. Reflect on the following questions, in the context of your unnecessary suffering:

> What kind of emotional legacy do I want to leave behind?
>
> How can I use my experiences to inspire hope in others?
>
> What practices can I commit to that reinforce this legacy?

Write down your statement, and revisit it regularly. Let it serve as a compass, guiding your choices and reminding you of the impact your journey can have beyond yourself.

¤ ¤ ¤ ¤ ¤

As we end our exploration of suffering, I offer you a simple yet profound truth:

The capacity to suffer less and to experience greater joy is within you.

Your brain, with all its complexities and quirks, is not your enemy; it is your most powerful ally when understood and skillfully guided.

Hope is not found outside of you; it is discovered within the sanctuary of your own mind. It is the quiet whisper that says, "You can," even when fear shouts, "You can't!" It is the steady pulse of your life force, resilient and persistent. To cultivate hope is to engage in a lifelong practice of mindful awareness, cognitive clarity, and emotional courage. It requires us to notice when fear-based narratives hijack our inner dialogue and to choose, again and again, to anchor ourselves in gratitude, curiosity, and trust.

This practice is neither easy nor passive. It demands that we show up for ourselves daily, that we remain present with our discomfort, and that we commit to nurturing the belief that joy is possible—even amidst life's inevitable challenges.

Janet's actions are a lesson to us to be both diligent and patient, as over time she noticed a shift in her perspective. Hope was also patient. She walked slowly and carefully with Janet, trusting that healing would arrive. We trusted her faith, and it bore fruit, in abundance.

Hope is not a wishful fantasy; it is a neurobiological reality that we can access when we engage the full spectrum of our cognitive, emotional, and physiological resources. It is the ember that glows quietly within us, waiting to be fanned into flame by our attention and care.

So, as you move forward from these pages and into your own life, I leave you with this benediction:

Hope always exists, even in darkness,
for shadows can only exist in the presence of light.
May you be defined not by your shadows
but by the light that lives within you.
And most of all, may you find,
through the words and ideas shared in this book,
the unwavering truth that suffering is optional
and hope is eternal.

ILLUSTRATOR'S NOTE ABOUT THE COVER

"Suffering that can be given meaning through growth is something we have reason to want. Suffering that remains brute and uninterpreted is something we have reason not to want."[134]

These are Christopher Janaway's observations on the words of well-known German philosopher Friedrich Nietzsche.

Nietzsche acknowledges that suffering promotes growth, making it crucial to our well-being. It contributes value to a person, and he celebrates it as meritorious: "If we value the overcoming of resistance, then we must also value the resistance that is an ingredient of it."[135]

With new eyes a person may look at suffering as more than torment; they may see grit, tenacity, and strength.

The cover art depicts the cycle of suffering and its relationship with hope. The cycle splits as hope intervenes. In this way suffering inherits its reason to be wanted, and growth transpires.

The flower is a symbol for the collection and transference of knowledge. It appears in a moment when knowledge is imparted or gained, a delicate reminder of growth. Flora adorns the interlocking hands as the flower nurses suffering into hope. A core characteristic of suffering is its transferable nature, how it persists in the mind and migrates from one person to the next. The hands grasp at one another, indicating these bonds laminated by suffering, both valuable and poisonous. The vines and hands are subtly suffocating, desperate, urgent...but they are also connective, strong, hopeful, and compassionate.

As we learn in this book, suffering is not just a journey of hurt and desperation; it is alive with sensitivity and necessity. It creates texture and tests humanity, a reminder of the indomitable human spirit.

Hannah Mae Silcock

CONNECT

You can visit me at my website, www.RoddyCarter.com. There, you will find more of my work that is designed to help you on your journey of conscious living, along with my recommendations for additional reading and research. Please reach out to me directly at connect@RoddyCarter.com to share your victories and struggles.

If this book resonated with you, I'd be honored if you'd leave a review on the site where you purchased it...your voice helps others find it.

FURTHER READING

Throughout history, humans have sought to understand and cope with the experience of suffering. Across cultures and spiritual traditions, unique insights have emerged, offering perspectives on the origins of suffering and pathways to transcend it. Whether envisioning suffering as a test, a purification process, or an opportunity for growth, these spiritual perspectives provide a rich and timeless tapestry of wisdom. They complement modern psychological and neurological understandings, reminding us that the struggle with and response to suffering is a universal theme woven into the fabric of the human condition. The following are some suggested readings for exploring spiritual perspectives on suffering.

Buddhism: *What the Buddha Taught*, Walpola Rahula

> In Buddhism, suffering (*dukkha*) is recognized as an inevitable aspect of existence, rooted in craving, attachment, and ignorance. The Buddha's Four Noble Truths outline suffering's nature, its causes, and a pathway to liberation. This perspective emphasizes that, while external circumstances may spark suffering, its perpetuation lies within the mind. The Eightfold Path, a practical guide for ending suffering, advocates mindfulness, ethical living, and cultivating awareness. These principles align closely with modern practices of mindfulness and emotional regulation, emphasizing the power of conscious thought in reducing suffering. As the *Dhammapada* reminds us, "All experiences are preceded by mind, led by mind, made by mind."[136]

Christianity: *The Problem of Pain*, C. S. Lewis

> Christianity acknowledges suffering as intrinsic to the human condition, often interpreting it as a test of faith or a means of spiritual purification. Jesus Christ's teachings encourage enduring suffering with faith, viewing it as a path to redemption and greater connection with God. Compassion, forgiveness, and hope are central tenets, offering solace amidst hardship. Through Christ's messages of love and forgiveness, suffering is reframed not as punishment but as an opportunity to draw closer to God and cultivate resilience.

Hinduism: *Bhagavad Gita: A New Translation*, Stephen Mitchell

In Hindu philosophy, suffering is woven into the fabric of *samsara*—the cycle of birth, death, and rebirth. This cycle, influenced by karma, represents the consequences of one's actions and shapes current and future experiences. However, suffering is also seen as an opportunity for spiritual growth and self-realization. Through practices like meditation (*dhyana*), devotion (*bhakti*), and selfless action (*karma yoga*), individuals can transcend their suffering and achieve *moksha*—liberation from the cycle of pain. The *Bhagavad Gita*, a cornerstone of Hindu thought, teaches detachment from the ego and outcomes, offering a framework for understanding life's challenges and moving toward inner peace.

Islam: *The Quran: A New Translation*, M. A. S. Abdel Haleem

In Islam, suffering is understood as a test from Allah, designed to purify and strengthen faith. Life is viewed as a temporary abode where trials serve to bring individuals closer to God. The virtues of patience (*sabr*) and gratitude (*shukr*) are emphasized as tools to navigate suffering, encouraging acceptance of trials as part of Allah's divine plan. Islamic teachings advocate resilience through prayer and community, grounding individuals in faith during times of hardship. The Quran, particularly verses in Surah Al-Baqarah (2:155–157), reassures believers that trials are opportunities to grow spiritually and that steadfastness will be rewarded.

Judaism: The Book of Job from the *Tanakh*

In Judaism, suffering is a multifaceted experience, often understood as a way to grow closer to God or refine one's character. It is also seen as part of a divine mystery that may not always be fully understood. The Jewish tradition encourages questioning and wrestling with suffering, exemplified in the story of Job and the lamentations of the Psalms. Practices like prayer, study, and acts of kindness offer pathways to find purpose amidst pain. This approach emphasizes resilience and the importance of community in overcoming life's challenges, making suffering a communal as well as an individual experience.

Stoicism: *Meditations*, by Marcus Aurelius

Though not a religion, Stoicism offers helpful insights into the nature of suffering, emphasizing that it arises not from external events but from our perceptions of them. Stoics advocate cultivating inner tranquility through rational thinking, self-discipline, and acceptance. Marcus Aurelius, in his *Meditations*, encourages detachment from things outside our control and reframing suffering as an opportunity for growth. This perspective resonates with modern cognitive-behavioral practices, reminding us that, while we cannot always change our circumstances, we can change how we respond to them.

Taoism: *Tao Te Ching*, Laozi (translated by Stephen Mitchell)

Taoism views suffering as arising from resistance to the natural flow of life (*Tao*). It is seen as a disturbance in the balance of existence, often rooted in desires and attachments. By living in harmony with the Tao, individuals can minimize suffering and cultivate peace. Practices like meditation, tai chi, and embracing simplicity encourage acceptance of life's challenges and detachment from rigid expectations. The *Tao Te Ching* by Laozi teaches that surrendering to life's natural rhythms allows individuals to find serenity amidst turbulence, a message that remains universally applicable.

While these spiritual perspectives differ in their approaches, they share a common emphasis on finding meaning, resilience, and a sense of control over one's internal experience. They complement modern understandings of suffering, offering tools and insights for navigating life's inevitable challenges. As we explore the roots and consequences of suffering, these timeless teachings remind us of humanity's enduring quest to transcend pain and embrace peace.

NOTES

1 Jung, C. J. (1980). *Psychology and alchemy.* (G. Adler & R. F. C. Hull, Trans.). Princeton University Press.

2 Batty, G. D., Russ, T. C., Stamatakis, E., & Kivimäki, M. (2014). Psychological distress and risk of peripheral vascular disease, abdominal aortic aneurysm, and heart failure: Pooling of sixteen cohort studies. *Atherosclerosis, 236*(2), 385–388. https://www.atherosclerosis-journal.com/article/S0021-9150(14)01255-6/abstract

3 Satyjeet, F., Naz, S., Kumar, V., Aung, N. H., Bansari, K., Irfan, S., & Rizwan, A. (2020). Psychological stress as a risk factor for cardiovascular disease: A case-control study. *Cureus, 12*(10), e10757. https://doi.org/10.7759/cureus.10757

4 Kawachi, I., Kyriopoulos, I., & Vandoros, S. (2023). Economic uncertainty and cardiovascular disease mortality. *Health Economics, 32*(7), 1550–1560. https://doi.org/10.1002/hec.4678

5 Kwapong, Y. A., Boakye, E., Khan, S. S., Honigberg, M. C., Martin, S. S., Oyeka, C. P., Hays, A. G., Natarajan, P., Mamas, M. A., Blumentahl, R. S., Blaha, M. J., & Sharma, G. (2023). Association of depression and poor mental health with cardiovascular disease and suboptimal cardiovascular health among young adults in the United States. *Journal of the American Heart Association, 12*(3), e028332. https://doi.org/10.1161/JAHA.122.028332

6 Reynolds, C. A., & Minic, Z. (2023). Chronic pain-associated cardiovascular disease: The role of sympathetic nerve activity. *International Journal of Molecular Sciences, 24*(6), 5378. https://doi.org/10.3390/ijms24065378

7 Munir, L. Z., & du Toit, E. F. (2024). Impact of chronic psychological stress on cardiovascular disease risk: A narrative review. *Heart and Mind, 8*(4), 268–278. https://doi.org/10.4103/hm.HM-D-24-00040

8 Lee, S. P., Sung, I., Kim, J. H., Lee, S., Park, H. S., & Shim, C. S. (2015). The effect of emotional stress and depression on the prevalence of digestive diseases. *Journal of Neurogastroenterology and Motility, 21*(2), 273–282. https://doi.org/10.5056/jnm14116

9 Edman, J. S., Greeson, J. M., Roberts, R. S., Kaufman, A. B., Abrams, D. I., Dolor, R. J., & Wolever, R. Q. (2017). Perceived stress in patients with common gastrointestinal disorders: Associations with quality of life, symptoms and disease management. *EXPLORE, 13*(2), 124–128. https://doi.org/10.1016/j.explore.2016.12.005

10 Labanski, A., Langhorst, J., Engler, H., & Elsenbruch, S. (2020). Stress and the brain-gut axis in functional and chronic-inflammatory gastrointestinal diseases: A transdisciplinary challenge. *Psychoneuroendocrinology, 111*, 104501. https://doi.org/10.1016/j.psyneuen.2019.104501

11 Agusti, A., Lamers, F., Tamayo, M., Benito-Amat, C., Molina-Mendoza, G. V., Penninx, B. W. J. H., & Sanz, Y. (2023). The gut microbiome in early life stress: A systematic review. *Nutrients, 15*(11), 2566. https://doi.org/10.3390/nu15112566

12 Hammarström, P., Rosendahl, S., Gruber, M., & Nordin, S. (2023). Somatic symptoms in burnout in a general adult population. *Journal of Psychosomatic Research, 168*, 111217. https://doi.org/10.1016/j.jpsychores.2023.111217

13 Leigh, S.-J., Uhlig, F., Wilmes, L., Sanchez-Diaz, P., Gheorghe, C. E., Goodson, M. S., Kelley-Loughnane, N., Hyland, N. P., Cryan, J. F., & Clarke, G. (2023). The impact of acute and chronic stress on gastrointestinal physiology and function: A microbiota–gut–brain axis perspective. *Journal of Physiology, 601*, 4491–4538. https://doi.org/10.1113/JP281951

14 Minjoz, S., Sinniger, V., Hot, P., Bonaz, B., & Pellissier, S. (2023). The burden of early life stress in chronic inflammatory bowel diseases. *Journal of Health Psychology, 28*(13), 1204–1216. https://doi.org/10.1177/13591053231173918

15 Morey, J. N., Boggero, I. A., Scott, A. B., & Segerstrom, S. C. (2015). Current directions in stress and human immune function. *Current Opinions in Psychology, 1*(5), 13–17. https://doi.org/10.1016/j.copsyc.2015.03.007

16 Song, H., Fang, F., Tomasson, G., Arnberg, F. K., Mataix-Cols, D., Fernández de la Cruz, L., Almqvist, C., Fall, K., & Valdimarsdóttir, A. (2018). Association of stress-related disorders with subsequent autoimmune disease. *JAMA, 319*(23), 2388–2400. https://jamanetwork.com/journals/jama/fullarticle/2685155

17 Bae, Y.-S., Shin, E.-C., Bae, Y.-S., & Van Eden, W. (2019). Editorial: Stress and immunity. *Frontiers in Immunology, 10*, 245. https://doi.org/10.3389/fimmu.2019.00245

18 Seiler, A., Fagundes, C. P., & Christian, L. M. (2020). The impact of everyday stressors on the immune system and health. In A. Choukèr (Ed.), *Stress challenges and immunity in space* (pp. 71–92). Springer.

19 Haykin, H., & Rolls, A. (2021). The neuroimmune response during stress: A physiological perspective. *Immunity, 54*(9), 1933–1947. https://doi.org/10.1016/j.immuni.2021.08.023

20 Alotiby, A. (2024). Immunology of stress: A review article. *Journal of Clinical Medicine, 13*(21), 6394. https://doi.org/10.3390/jcm13216394

21 Reme, S. (2016). Anxiety could play a larger role than depression in migraine headache. *Scandinavian Journal of Pain, 13*(1), 127–127. https://doi.org/10.1016/j.sjpain.2016.08.009

22 Buscemi, V., Chang, W. J., Liston, M.B., McAuley, J. H., & Schabrun, S. (2017). The role of psychosocial stress in the development of chronic musculoskeletal pain disorders: Protocol for a systematic review and meta-analysis. *Systematic Reviews, 6*, 224. https://doi.org/10.1186/s13643-017-0618-0

23 Christensen, J. O., Johansen, S., & Knardahl, S. (2017). Psychological predictors of change in the number of musculoskeletal pain sites among Norwegian employees: A prospective study. *BMC Musculoskeletal Disorders, 18*, 140. https://doi.org/10.1186/s12891-017-1503-7

24 Alsaadi, S. M. (2022). Musculoskeletal pain in undergraduate students is significantly associated with psychological distress and poor sleep quality. *International Journal of Environmental Research and Public Health, 19*(21), 13929. https://doi.org/10.3390/ijerph192113929

25 Goksan Yavuz, B., Acar, E., Sancak, B., Sayin, E., Yalinay Dikmen, P., & Ilgaz Aydinlar, E. (2022). The role of metacognition, negative automatic thoughts and emotions in migraine-related disability among adult migraine patients. *Psychology, Health & Medicine, 28*(10), 3177–3189. https://doi.org/10.1080/13548506.2022.2104884

26 Ruela, G. de A., Barreto, S. M., Griep, R. H., Benseñor, I. M., Telles, R. W., & Camelo, L. V. (2022). Job stress and chronic and widespread musculoskeletal pain: A cross-sectional analysis from the Brazilian Longitudinal Study of Adult Health Musculoskeletal. *PAIN, 163*(10), 2044–2051. https://doi.org/10.1097/j.pain.0000000000002602

27 Collins, S., McDevitt-Petrovic, O., & Henderson, E. (2023). Can emotional distress induce acute low back pain? A systematic review. *Physiotherapy Practice and Review, 44*(2), 177–186. https://doi.org/10.3233/PPR-230737

28 Yap, A. U., & Marpaung, C. (2023). Correlates between temporomandibular disorder severity, emotional distress, and eudaimonic well-being among young adults. *Oral Diseases, 29*(7), 2780–2788. https://doi.org/10.1111/odi.14343

29 Gustavson, D. E., du Pont, A., Whisman, M. A., & Miyake, A. (2018). Evidence for transdiagnostic repetitive negative thinking and its association with rumination, worry, and depression and anxiety symptoms: A commonality analysis. *Collabra Psychology, 4*(1), 13. https://doi.org/10.1525/collabra.128

30 LeMoult, J. (2020). From stress to depression: Bringing together cognitive and biological science. *Current Directions in Psychological Science, 29*(6), 592–598. https://doi.org/10.1177/0963721420964039

31 Akil, H., & Nestler, E. J. (2023). The neurobiology of stress: Vulnerability, resilience, and major depression. *Proceedings of the National Academy of Sciences U.S.A., 120*(49), e2312662120. https://doi.org/10.1073/pnas.2312662120

32 LeDuke, D. O., Borio, M., Miranda, R., & Tye, K. M. (2023). Anxiety and depression: A top-down, bottom-up model of circuit function. *Annals of the New York Academy of Sciences, 1525*(1), 70–87. https://doi.org/10.1111/nyas.14997

33 Mohammadkhani, S., Attar, F., & Akbari, M. (2023). The linkage between negative affectivity with emotional distress in college student: The mediator and moderator role of difficulty in emotion regulation, repetitive negative thinking, and emotional invalidation. *Current Psychology, 42*, 6216–6228. https://doi.org/10.1007/s12144-021-01904-3

34 Sirois, F. M. (2023). Procrastination and stress: A conceptual review of why context matters. *International Journal of Environmental Research and Public Health, 20*(6), 5031. https://doi.org/10.3390/ijerph20065031

35 Zhang, L., Kong, M., Li, Z., Zhao, X., & Gao, L. (2018). Chronic stress and moral decision-making: An exploration with the CNI model. *Frontiers in Psychology, 9*. https://doi.org/10.3389/fpsyg.2018.01702

36 Wu, J., Feng, M., Liu, Y., Fang, H., & Duan, H. (2019). The relationship between chronic perceived stress and error processing: Evidence from event-related potentials. *Scientific Reports, 9*, 11605. https://doi.org/10.1038/s41598-019-48179-0

37 Liu, Q., Liu, Y., Leng, X., Han, J., Xia, F., & Chen, H. (2020). Impact of chronic stress on attention control: Evidence from behavioral and event-related potential analyses. *Neuroscience Bulletin, 36*(11), 1395–1410. https://doi.org/10.1007/s12264-020-00549-9

38 Vartanian, O., Saint, S. A., Herz, N., & Suedfeld, P. (2020). The creative brain under stress: Considerations for performance in extreme environments. *Frontiers in Psychology, 11*, 585969. https://doi.org/10.3389/fpsyg.2020.585969

39 Almarzouki, A. F. (2024). Stress, working memory, and academic performance: A neuroscience perspective. *Stress, 27*(1). https://doi.org/10.1080/10253890.2024.2364333

40 Leggett, A., Burgard, S., & Zivin, K. (2016). The impact of sleep disturbance on the association between stressful life events and depressive symptoms. *The Journals of Gerontology: Series B, 71*(1), 118–128. https://doi.org/10.1093/geronb/gbv072

41 Kalmbach, D. A., Anderson, J. R., & Drake, C. L. (2018). The impact of stress on sleep: Pathogenic sleep reactivity as a vulnerability to insomnia and circadian disorders. *Journal of Sleep Research, 27*(6), e12710. https://doi.org/10.1111/jsr.12710

42 Li, Y., Gu, S., Wang, Z., Li, H., Xu, X., Zhu, H., Deng, S., Ma, X., Feng, G., Wang, F., & Huang, J. H. (2019). Relationship between stressful life events and sleep quality: Rumination as a mediator and resilience as a moderator. *Frontiers in Psychiatry, 10*, 348. https://doi.org/10.3389/fpsyt.2019.00348

43 Sladek, M. R., Doane, L. D., & Breitenstein, R. S. (2019). Daily rumination about stress, sleep, and diurnal cortisol activity. *Cognition and Emotion, 34*(2), 188–200. https://doi.org/10.1080/02699931.2019.1601617

44 Yap, Y., Slavish, D. C., Taylor, D. J., Bei, B., & Wiley, J. F. (2020). Bi-directional relations between stress and self-reported and actigraphy-assessed sleep: A daily intensive longitudinal study. *Sleep, 43*(3), zsz250. https://doi.org/10.1093/sleep/zsz250

45 Merrill, R. M. (2022). Mental health conditions according to stress and sleep disorders. *International Journal of Environmental Research and Public Health, 19*(13), 7957. https://doi.org/10.3390/ijerph19137957

46 Bunmi, O., Olatunji, K. A., Knowles, R. C., Cox, D., & Cole, A. (2023). Linking repetitive negative thinking and insomnia symptoms: A longitudinal trait-state model. *Journal of Anxiety Disorders, 97*, 102732. https://doi.org/10.1016/j.janxdis.2023.102732

47 Kuo, W.-C., Ersig, A. L., Kunkul, F., Brown, R. L., & Oakley, L. D. (2023). Linking chronic stress to insomnia symptoms in older adults: The role of stress co-occurrence during the pandemic. *Research in Nursing & Health, 46*(1), 68–79. https://doi.org/10.1002/nur.22278

48 Sullivan, E. C., James, E., Henderson, L.-M., McCall, C., & Cairney, S. A. (2023). The influence of emotion regulation strategies and sleep quality on depression and anxiety. *Cortex, 166*, 286–305. https://doi.org/10.1016/j.cortex.2023.06.001

49 Tomova, L., von Dawans, B., Heinrichs, M., Silani, G., & Lamm, C. (2014). Is stress affecting our ability to tune into others? Evidence for gender differences in the effects of stress on self-other distinction. *Psychoneuroendocrinology, 43*, 95–104. https://doi.org/10.1016/j.psyneuen.2014.02.006

50 Eckhardt, C. I., & Parrott, D. J. (2017). Stress and intimate partner aggression. *Current Opinion in Psychology, 13*, 153–157. https://doi.org/10.1016/j.copsyc.2016.09.005

51 Durham, T. A., Byllesby, B. M., Lv, X., Elhai, J. D., & Wang, L. (2018). Anger as an underlying dimension of posttraumatic stress disorder. *Psychiatry Research, 267*, 535–540. https://doi.org/10.1016/j.psychres.2018.06.011

52 Neff, L. A., Nguyen, T. T. T., & Williamson, H. C. (2021). Too stressed to help? The effects of stress on noticing partner needs and enacting support. *Personality and Social Psychology Bulletin, 47*(11), 1565–1579. https://doi.org/10.1177/0146167220974490

53 Galanakis, M., Kallianta, M., Katsira, C., Liakopoulou, D., Chrousos, P., & Darviri, C. (2015). The association between stress and sexual dysfunctionality in men and women: A systematic review. *Psychology, 6*(14), 1888–1892. https://doi.org/10.4236/psych.2015.614186

54 Randall, A. K., & Bodenmann, G. (2017). Stress and its associations with relationship satisfaction. *Current Opinion in Psychology, 13*, 96–106. https://doi.org/10.1016/j.copsyc.2016.05.010

55 Sucrese, A. M., Neff, L. A., & Gleason, M. E. J. (2024). "Not now, I am too stressed": Stress and physical intimacy in early marriage. *Personality and Social Psychology Bulletin*. https://doi.org/10.1177/01461672241239134

56 Santini, Z. I., Jose, P. E., Cornwell, E. Y., Koyanagi, A., Nielsen, L., Hinrichsen, C., Meilstrup, C., Madsen, K. R., & Koushede, V. (2020). Social disconnectedness, perceived isolation, and symptoms of depression and anxiety among older Americans (NSHAP): A longitudinal mediation analysis. *The Lancet Public Health, 5*(1), e62–e70. https://www.thelancet.com/journals/lanpub/article/PIIS2468-2667(19)30230-0/fulltext

57 Emerson, K., Boggero, I., Ostir, G., & Jayawardhana, J. (2018). Pain as a risk factor for loneliness among older adults. *Journal of Aging and Health, 30*(9), 1450–1461. https://doi.org/10.1177/0898264317721348

58 Karayannis, N. V., Baumann, I., Sturgeon, J. A., Melloh, M., & Mackey, S. C. (2019). The impact of social isolation on pain interference: A longitudinal study. *Annals of Behavioral Medicine, 53*(1), 65–74. https://doi.org/10.1093/abm/kay017

59 Allen, S. F., Gilbody, S., Atkin, K., & van der Feltz-Cornelis, C. (2020). The associations between loneliness, social exclusion and pain in the general population: A N=502,528 cross-sectional UK Biobank study. *Journal of Psychiatric Research, 130*, 68–74. https://doi.org/10.1016/j.jpsychres.2020.06.028

60 Loeffler, A., & Steptoe, A. (2021). Bidirectional longitudinal associations between loneliness and pain, and the role of inflammation. *PAIN, 162*(3), 930–937. https://doi.org/10.1097/j.pain.0000000000002082

61 Bai, R. Impact of financial literacy, mental budgeting and self control on financial wellbeing: Mediating impact of investment decision making. *PLoS ONE, 18*(11), e0294466. https://doi.org/10.1371/journal.pone.0294466

62 Ami, R. B. (2023). Economic decision-making and mental health. *SSRN.* https://dx.doi.org/10.2139/ssrn.4363934

63 Takahashi, T., Oono, H., Inoue, T., Boku, S., Kako, Y., Kitaichi, Y., Kusumi, I., Masui, T., Nakagawa, S., Suzuki, K., Tanak, T., Koyama, T., & Radford, M. H. B. (2008). Depressive patients are more impulsive and inconsistent in intertemporal choice behavior for monetary gain and loss than healthy subject - an analysis based on Tsallis' statistics. *Neuro Endocrinology Letters, 29*(3), 351–358. https://arxiv.org/abs/1111.6493

64 Ettman, C. K., Thornburg, B., Abdalla, S. M., Meiselbach, M. K., & Galea, S. (2024). Financial assets and mental health over time. *Scientific Reports, 14*, 27370. https://www.nature.com/articles/s41598-024-76990-x

65 Bliss, T., Collingridge, G., Kaang, B. K., & Zhuo, M. (2016). Synaptic plasticity in the anterior cingulate cortex in acute and chronic pain. *Nature Reviews Neuroscience, 17*, 485–496. https://doi.org/10.1038/nrn.2016.68

66 Jee, H. J., Zhu, E., Sun, M., Liu, W., Zhang, Q., & Wang, J. (2023). Anterior cingulate cortex regulates pain catastrophizing-like behaviors in rats. *Molecular Brain, 16*, 71. https://doi.org/10.1186/s13041-023-01060-8

67 Groh, A., Mease, R., & Krieger, P. (2017). Pain processing in the thalamocortical system. *e-Neuroforum, 23*(3), 117–122. https://doi.org/10.1515/nf-2017-A019

68 Thompson, J. M., & Neugebauer, V. (2019). Cortico-limbic pain mechanisms. *Neuroscience Letters, 29*(702), 15–23. https://doi.org/10.1016/j.neulet.2018.11.037

69 Neugebauer, V. (2015). Amygdala pain mechanisms. In M. C. Michel's (Ed.), *Handbook of experimental pharmacology* (vol. 227, pp. 261–284). Springer.

70 Seno, M. D. J., Assis, D. V., Gouveia, F., Antunes, G. F., Kuroki, M., Oliveira, C. C., Santos, L. C. T., Pagano, R. L., & Martinez, R. C. R. (2018). The critical role of amygdala subnuclei in nociceptive and depressive-like behaviors in peripheral neuropathy. *Scientific Reports, 8*, 13608. https://doi.org/10.1038/s41598-018-31962-w

71 Elman, I., & Borsook, D. (2018). Threat response system: Parallel brain processes in pain vis-à-vis fear and anxiety. *Frontiers in Psychiatry, 9*(29). https://doi.org/10.3389/fpsyt.2018.00029

72 McCarberg, B., & Peppin, J. (2019). Pain pathways and nervous system plasticity: Learning and memory in pain. *Pain Medicine, 20*(12), 2421–2437. https://doi.org/10.1093/pm/pnz017

73 Veinante, P., Yalcin, I., & Barrot, M. (2013). The amygdala between sensation and affect: A role in pain. *Journal of Molecular Psychiatry, 1*(9). https://doi.org/10.1186/2049-9256-1-9

74 Neugebauer, V. (2015). Amygdala pain mechanisms. In M. C. Michel's (Ed.), *Handbook of experimental pharmacology* (vol. 227, pp. 261–284). Springer.

75 Ibid.

76 Veinante, P., Yalcin, I., & Barrot, M. (2013). The amygdala between sensation and affect: A role in pain. *Journal of Molecular Psychiatry, 1*(9). https://doi.org/10.1186/2049-9256-1-9

77 Yang, Y., & Wang, J.-Z. (2017). From structure to behavior in basolateral amygdala-hippocampus circuits. *Frontiers in Neural Circuits, 11*, 86. https://doi.org/10.3389/fncir.2017.00086

78 Costa, M., Lozano-Soldevilla, D., Gil-Nagel, A., Toledano, R., Oehrn, C. R., Kunz, L., Yebra, M., Mendez-Bertolo, C., Stieglitz, L., Sarnthein, J., Axmarcher, N., Moratti, S., & Strange, B. A. (2022). Aversive memory formation in humans involves an amygdala-hippocampus phase code. *Nature Communications, 13*, 6403. https://doi.org/10.1038/s41467-022-33828-2

79 Meerw jk, E. L., Ford, J. M., & Weiss, S. J. (2013). Brain regions associated with psychological pain: Implications for a neural network and its relationship to physical pain. *Brain Imaging and Behavior, 7*(1), 1–14. https://doi.org/10.1007/s11682-012-9179-y

80 Abdallah, C. G., & Geha, P. (2017). Chronic pain and chronic stress: Two sides of the same coin? *Chronic Stress, 1*. https://doi.org/10.1177/2470547017704763

81 Neugebauer, V. (2020). Amygdala physiology in pain. In M. C. Michel's (Ed.), *Handbook of behavioral neuroscience* (vol. 26, pp. 101–113). Springer.

82 Becker, L. J., Fillinger, C., Waegaert, R., Journée, S. H., Hener, P., Ayazgok, B., Humo, M., Karatas, M., Thouaye, M., Gaikwad, M., Degiorgis, L., Santin, M. d. N., Mondino, M., Barrot, M., Ibrahim, E. C., Turecki, G., Belzeaux, R., Veinante, P., Harsan, L. A.,...Yalcin, I. (2023). The basolateral amygdala-anterior cingulate pathway contributes to depression-like behaviors and comorbidity with chronic pain behaviors in male mice. *Nature Communications, 14*, 2198. https://doi.org/10.1038/s41467-023-37878-y

83 McDonald, A. J., & Mott, D. D. (2017). Functional neuroanatomy of amygdalohippocampal interconnections and their role in learning and memory. *Journal of Neuroscience Research, 95*(3), 797–820. https://doi.org/10.1002/jnr.23709

84 Thompson, J. M., & Neugebauer, V. (2017). Amygdala plasticity and pain. *Pain Research and Management*, 8296501. https://doi.org/10.1155/2017/8296501

85 De Ridder, D., Adhia, D., & Vanneste, S. (2021). The anatomy of pain and suffering in the brain and its clinical implications. *Neuroscience & Biobehavioral Reviews, 130*, 125–146. https://doi.org/10.1016/j.neubiorev.2021.08.013

86 Neugebauer, V. (2020). Amygdala physiology in pain. In M. C. Michel's (Ed.), *Handbook of behavioral neuroscience* (vol. 26, pp. 101–113). Springer.

87 Ong, W. Y., Stohler, C. S., & Herr, D. R. (2019). Role of the prefrontal cortex in pain processing. *Molecular Neurobiology, 56*, 1137–1166. https://doi.org/10.1007/s12035-018-1130-9

88 Yang, S., & Chang, M. C. (2019). Chronic pain: Structural and functional changes in brain structures and associated negative affective states. *International Journal of Molecular Sciences, 20*(13), 3130. https://doi.org/10.3390/ijms20133130

89 Neugebauer, V. (2015). Amygdala pain mechanisms. In M. C. Michel's (Ed.), *Handbook of experimental pharmacology* (vol. 227, pp. 261–284). Springer.

90 Wood, K. H., Ver Hoef, L. W., & Knight, D. C. (2014). The amygdala mediates the emotional modulation of threat-elicited skin conductance response. *Emotion, 14*(4), 693–700. https://doi.org/10.1037/a0036636

91 Lapate, R. C., Rokers, B., Tromp, D. P. M., Orfali, N. S., Oler, J. A., Doran, S. T., Adluru, N., Alexander, A. L., & Davidson, R. J. (2016). Awareness of emotional stimuli determines the behavioral consequences of amygdala activation and amygdala-prefrontal connectivity. *Scientific Reports, 6*, 25826. https://doi.org/10.1038/srep25826

92 Feller, L., Feller, G., Ballyram, T., Chandran, R., Lemmer, J., & Khammissa, R. A. G. (2020). Interrelations between pain, stress and executive functioning. *British Journal of Pain, 14*(3), 188–194. https://doi.org/10.1177/2049463719889380

93 Ressler, K. J. (2010). Amygdala activity, fear, and anxiety: Modulation by stress. *Biological Psychiatry, 67*(12), 1117–1119. https://doi.org/10.1016/j.biopsych.2010.04.027

94 Cai, Y.-Q., Wang, W., Paulucci-Holthauzen, A., & Pan, Z. Z. (2018). Brain circuits mediating opposing effects on emotion and pain. *Journal of Neuroscience, 38*(28), 6340–6349. https://doi.org/10.1523/JNEUROSCI.2780-17.2018

95 Puderbaugh, M., & Emmady, P. D. (2023). *Neuroplasticity*. StatPearls Publishing. https://pubmed.ncbi.nlm.nih.gov/32491743/

96 Neugebauer, V. (2015). Amygdala pain mechanisms. In M. C. Michel's (Ed.), *Handbook of experimental pharmacology* (vol. 227, pp. 261–284). Springer.

97 Cai, Y.-Q., Wang, W., Paulucci-Holthauzen, A., & Pan, Z. Z. (2018). Brain circuits mediating opposing effects on emotion and pain. *Journal of Neuroscience, 38*(28), 6340–6349. https://doi.org/10.1523/JNEUROSCI.2780-17.2018

98 Asanova, A., Ovdii, M., Mukharovska, I., & Skrynnyk, O. (2024). The role of catastrophizing, depression and anxiety in chronic pain: A cross-sectional pilot study. *International Neurological Journal, 20*(7), 334–342. https://doi.org/10.22141/2224-0713.20.7.2024.1113

99 Delgado-Gallén, S., Soler, M. D., Cabello-Toscano, M., Abellaneda-Pérez, K., Solana-Sánchez, J., España-Irla, G., Roca-Ventura, A., Bartrés-Faz, D., Tormos, J. M., Pascual-Leone, A., & Cattaneo, G. (2023). Brain system segregation and pain catastrophizing in chronic pain progression. *Frontiers in Neuroscience, 17*, 1148176. https://doi.org/10.3389/fnins.2023.1148176

100 Renna, M. E., Rosie Shrout, M., Madison, A. A., Lustberg, M., Povoski, S. P., Agnese, D. M., Reinbolt, R. E., Wesolowski, R., Williams, N. O., Ramaswamy, B., Sardesai, S. D., Noonan, A. M., VanDeusen, J. B., Malarkey, W. B., & Kiecolt-Glaser, J. K. (2021). Worry and rumination in breast cancer patients: Perseveration worsens self-rated health. *Journal of Behavioral Medicine, 44*, 253–259. https://doi.org/10.1007/s10865-020-00192-9

101 Gustavson, D. E., du Pont, A., Whisman, M. A., & Miyake, A. (2018). Evidence for transdiagnostic repetitive negative thinking and its association with rumination, worry, and depression and anxiety symptoms: A commonality analysis. *Collabra: Psychology, 4*(1), 13. https://doi.org/10.1525/collabra.128

102 Thompson, J. M., & Neugebauer, V. (2017). Amygdala plasticity and pain. *Pain Research and Management*, 8296501. https://doi.org/10.1155/2017/8296501

103 Yang, S., & Chang, M. C. (2019). Chronic pain: Structural and functional changes in brain structures and associated negative affective states. *International Journal of Molecular Sciences, 20*(13), 3130. https://doi.org/10.3390/ijms20133130

104 Amaro-Díaz, L., Montoro, C. I., Fischer-Jbali, L. R., & Galvez-Sánchez, C. M. (2022). Chronic pain and emotional stroop: A systematic review. *Journal of Clinical Medicine, 11*(12), 3259. https://doi.org/10.3390/jcm11123259

105 Costa, M., Lozano-Soldevilla, D., Gil-Nagel, A., Toledano, R., Oehrn, C. R., Kunz, L., Yebra, M., Mendez-Bertolo, C., Stieglitz, L., Sarnthein, J., Axmacher, N., Moratti, S., & Strange, B. A. (2022). Aversive memory formation in humans involves an amygdala-hippocampus phase code. *Nature Communications, 13*, 6403. https://doi.org/10.1038/s41467-022-33828-2

106 Qasim, S. E., Mohan, U. R., Stein, J. M., & Jacobs, J. (2023). Neuronal activity in the human amygdala and hippocampus enhances emotional memory encoding. *Natural Human Behavior, 7*(5), 754–764. https://doi.org/10.1038/s41562-022-01502-8

107 Sherman, B. E., Harris, B. B., Turk-Browne, N. B., Sinha, R., & Goldfarb, E. V. (2023). Hippocampal mechanisms support cortisol-induced memory enhancements. *Journal of Neuroscience, 43*(43), 7198–7212. https://doi.org/10.1523/JNEUROSCI.0916-23.2023

108 Wan, Z., Rolls, E. T., Feng, J., & Cheng, W. (2022). Brain functional connectivities that mediate the association between childhood traumatic events, and adult mental health and cognition. *eBioMedicine, 79*, 104002. https://doi.org/10.1016/j.ebiom.2022.104002

109 Cai, J., Li, J., Liu, D., Gao, S., Zhao, Y., Zhang, J., & Liu, Q. (2023). Long-term effects of childhood trauma subtypes on adult brain function. *Brain and Behavior, 13*(5), e2981. https://doi.org/10.1002/brb3.2981

110 Laricchiuta, D., Panuccio, A., Picerni, E., Biondo, D., Genovesi, B., & Petrosini, L. (2023). The body keeps the score: The neurobiological profile of traumatized adolescents. *Neuroscience & Biobehavioral Reviews, 145*, 105033. https://doi.org/10.1016/j.neubiorev.2023.105033

111 Madden, R. A., Atkinson, K., Shen, X., Green, C., Hillary, R. F., Hawkins, E., Såge, E., Sandu, A.-L., Waiter, G., McNeil, C., Harris, M., Campbell, A., Porteous, D., Macfarlane, J. A., Murray, A., Steele, D., Romaniuk, L., Lawrie, S. M., McIntosh, A. M., & Whalley, H. C. (2023). Structural brain correlates of childhood trauma with replication across two large, independent community-based samples. *European Psychiatry, 66*(1), e19. https://doi.org/10.1192/j.eurpsy.2022.2347

112 Peverill, M., Rosen, M. L., Lurie, L. A., Sambrook, K. A., Sheridan, M. A., & McLaughlin, K. A. (2023). Childhood trauma and brain structure in children and adolescents. *Developmental Cognitive Neuroscience, 59*, 101180. https://doi.org/10.1016/j.dcn.2022.101180

113 Orellana, S. C., Bethlehem, R. A. I., Simpson-Kent, I. L., van Harmelen, A., Vértes, P. E., & Bullmore, E. T. (2024). Childhood maltreatment influences adult brain structure through its effects on immune, metabolic, and psychosocial factors. *Proceedings of the National Academy of Sciences U.S.A., 121*(16), e2304704121. https://doi.org/10.1073/pnas.2304704121

114 Van der Kolk, B. A. (2014). *The body keeps the score: Brain, mind, and body in the healing of trauma*. Penguin.

115 Franco, F. (2023). Intergenerational transmission of trauma. *Journal of Health Service Psychology, 49*, 185–190. https://doi.org/10.1007/s42843-023-00096-7

116 Scorza, P., Duarte, C. S., Lee, S., Wu, H., Posner, J., Baccarelli, A., & Monk, C. (2023). Stage 2 registered report: Epigenetic intergenerational transmission: Mothers' adverse childhood experiences and DNA methylation. *Journal of the American Academy of Child and Adolescent Psychiatry, 62*(10),1110–1122. https://doi.org/10.1016/j.jaac.2023.02.018

117 Bozack, A. K., Merrill, S. M., & Cardenas, A. (2024). Epigenetic biomarkers for understanding adverse experiences and health. *JAMA Network Open, 7*(8), e2427070. https://doi.org/10.1001/jamanetworkopen.2024.27070

118 Holuka, C., Grova, N., Charalambous, E. G., Le Cléac, H, J., Turner, J. D., & Mposhi, A. (2024). Transgenerational impacts of early life adversity: From health determinants, implications to epigenetic consequences. *Neuroscience & Biobehavioral Reviews, 164*, 105785. https://doi.org/10.1016/j.neubiorev.2024.105785

119 Yehuda, R., & Lehrner, A. (2018). Intergenerational transmission of trauma effects: Putative role of epigenetic mechanisms. *World Psychiatry, 17*(3), 243–257. https://doi.org/10.1002/wps.20568

120 Dashorst, P., Mooren, T. M., Kleber, R. J., de Jong, P. J., & Huntjens, R. J. C. (2019). Intergenerational consequences of the Holocaust on offspring mental health: A systematic review of associated factors and mechanisms. *European Journal of Psychotraumatology, 10*(1), 1654065. https://doi.org/10.1080/20008198.2019.1654065

121 Ahmed, F. (2010). Epigenetics: Tales of adversity. *Nature, 468*, S20. https://doi.org/10.1038/468S20a

122 Zimmer, C. (2018, January 31). The famine ended 70 years ago but Dutch genes still bear scars. *The New York Times*. https://www.nytimes.com/2018/01/31/science/dutch-famine-genes.html

123 De Rooij, S. R., Bleker, L. S., Painter, R. C., Ravelli, A. C., & Roseboom, T. J. (2021). Lessons learned from 25 years of research into long term consequences of prenatal exposure to the Dutch famine 1944–45: The Dutch famine birth cohort. *International Journal of Environmental Health Research, 32*(7), 1432–1446. https://doi.org/10.1080/09603123.2021.1888894

124 Ramírez-Smith, D., & Palloni, A. (2023). You are what your parents ate: A study on the intergenerational effects of famine exposure on cardiometabolic health in later life. *Population Association of America's Annual Meeting*. http://hdl.handle.net/10261/366614

125 Costa, D. L. (2012). Scarring and mortality selection among Civil War POWs: A long-term mortality, morbidity, and socioeconomic follow-up. *Demography, 49*(4), 1185–1206. https://doi.org/10.1007/s13524-012-0125-9

126 Costa, D. L., Yetter, N., & DeSomer, H. (2018). Intergenerational transmission of paternal trauma among US Civil War ex-POWs. *Proceedings of the National Academy of Sciences U.S.A., 115*(44), 11215–11220. https://doi.org/10.1073/pnas.1803630115

127 Costa, D. L. (2023). Overweight grandsons and grandfathers' starvation exposure. *Journal of Health Economics, 91*, 102796. https://doi.org/10.1016/j.jhealeco.2023.102796

128 Patel, J. (2024). Advances in the study of mirror neurons and their impact on neuroscience: An editorial. Cureus, 16(5), e61299. https://pmc.ncbi.nlm.nih.gov/articles/PMC11212500/

129 Van der Kolk, B. A. (2014). *The body keeps the score: Brain, mind, and body in the healing of trauma*. Penguin.

130 Hyman, S. E. (2000). The genetics of mental illness: Implications for practice. *Bulletin of the World Health Organization, 78*(4), 455–463.

131 Van Houtum, L. A. E. M., Baaré, W. F. C., Beckmann, C. F., Castro-Fornieles, J., Cecil, C. A. M., Dittrich, J., Ebdrup, B. H., Fegert, J. M., Havdahl, A., Hillegers, M. H. J., Kalisch, R., Kushner, S. A., Mansuy, I. M., Mežinska, S., Moreno, C., Muetzel, R. L., Neumann, A., Nordentoft, M., Pingault, J.-B.,...van Haren, N. E. M. (2024) Running in the FAMILY: Understanding and predicting the intergenerational transmission of mental illness. *European Child & Adolescent Psychiatry, 33*, 3885–3898. https://link.springer.com/article/10.1007/s00787-024-02423-9

132 Perroud, N., Rutembesa, E., Paoloni-Giacobino, A., Mutabaruka, J., Mutesa, L., Stenz, L., Malafosse, A., & Karege, F. (2014). The Tutsi genocide and transgenerational transmission of maternal stress: Epigenetics and biology of the HPA axis. *World Journal of Biological Psychiatry, 15*(4), 334–345. https://pubmed.ncbi.nlm.nih.gov/24690014/

133 Center on the Developing Child. (2007). *In brief: The science of early childhood development.* Harvard University. https://developingchild.harvard.edu/resources/inbriefs/inbrief-science-of-ecd/

134 Janaway, C. (2017). Attitudes to suffering: Parfit and Nietzsche. *Inquiry, 60*(1-2), 66–95. https://doi.org/10.1080/0020174X.2016.1251165

135 Ibid.

136 *The Dhammapada: A new translation of the Buddhist classic with annotations* (G. Fronsdal, Trans.). (2005). Shambhala Publications.

For more information, visit www.RoddyCarter.com,

or reach out directly via connect@RoddyCarter.com.

www.ingramcontent.com/pod-product-compliance
Lightning Source LLC
LaVergne TN
LVHW071410210926
848486LV00004B/241